Annals of Mathematics Studies

Number 130

Harmonic Maps and Minimal Immersions with Symmetries

METHODS OF ORDINARY DIFFERENTIAL EQUATIONS
APPLIED TO ELLIPTIC VARIATIONAL PROBLEMS

by

James Eells and Andrea Ratto

PRINCETON UNIVERSITY PRESS

PRINCETON, NEW JERSEY
1993

Copyright © 1993 by Princeton University Press
ALL RIGHTS RESERVED

Printed in the United States of America

The Annals of Mathematics Studies are edited by
Luis A. Caffarelli, John N. Mather, and Elias M. Stein

Princeton University Press books are printed on acid-free paper,
and meet the guidelines for permanence and durability of the
Committee on Production Guidelines for Book Longevity of the Council
on Library Resources

Library of Congress Catalog-in-Publication Data

Eells, James, 1926-
Harmonic maps and minimal immersions with symmetries : methods of
ordinary differential equations applied to elliptic variational problems /
by James Eells and Andrea Ratto.
p. cm. — (Annals of mathematics studies ; no. 130)
Includes bibliographical references and index.
ISBN 0-691-03321-8 —ISBN 0-691-10249-X (pbk.)
1. Harmonic maps. 2. Immersions (Mathematics). 3. Differential
equations, Elliptic—Numerical solutions. I. Ratto, Andrea, 1961-.
QA614.73.E35 1993
514'.7—dc20   92-38760

Harmonic Maps and Minimal Immersions with Symmetries

HARMONIC MAPS AND MINIMAL IMMERSIONS
WITH SYMMETRIES

Methods of ordinary differential equations
applied to elliptic variational problems

James Eells and Andrea Ratto

INTRODUCTION

In this monograph we study harmonic maps, minimal and parallel mean curvature immersions in various symmetric contexts. The maps under consideration are solutions to certain elliptic variational problems; and in the presence of suitable symmetry, those often admit reductions to lower dimensional problems whose qualitative study is more manageable.

(1) Our reduction theory (Chapter IV) is based on the differential geometry of fibre bundles - just as our variational theory is expressed in terms of the calculus in Riemannian vector bundles.

For a start: A variation $(\varphi_t)_{t \in \mathbb{R}}$ of a map $\varphi_0 = \varphi : M \to N$ between two Riemannian manifolds determines a section

$$v = \left.\frac{\partial \varphi_t}{\partial t}\right|_{t=0}$$

of the induced vector bundle $\varphi^{-1}T(N) \to M$ (defined in Chapter I (1.1)). We can interpret the differential $d\varphi : T(M) \to T(N)$ as a section of the tensor product $T^*(M) \otimes \varphi^{-1}T(N) \to M$, where $T^*(M)$ denotes the cotangent bundle of M. All of these vector bundles have Riemannian structures (i.e., a metric a on the fibres, together with a linear connection ∇ for which $\nabla a = 0$) induced from the metrics of M and N.

The *second fundamental* form of φ is the section of $\odot^2 T^*(M) \otimes \varphi^{-1}T(N)$ given by

(2) $$(\nabla d\varphi)(X,Y) = \nabla^\varphi_X(d\varphi.Y) - d\varphi(\nabla^M_X Y),$$

where ∇^M denotes the Levi-Civita connection of M and ∇^φ the connection on $\varphi^{-1}T(N)$ induced from ∇^N.

(3) The energy of a map $\varphi : M \to N$ is

$$E(\varphi) = \frac{1}{2}\int_M |d\varphi|^2 \, dx,$$

where dx denotes the volume element of the metric of M. Great care must be exercised in application of general variational theory. For instance,
i) E does not achieve its minimum on every homotopy class H. Indeed, if $\pi_1(N) = 0$ and $\pi_2(N) = 0$, then $\text{Inf}\{E(\varphi) : \varphi \in H\} = 0$. See also Appendix 1.
ii) Most of the standard *a priori* estimates of elliptic variational theory require that the image $\varphi(M)$ lie in a sufficiently small geodesic ball of N.

Assuming M compact (for simplicity of exposition),

(4) $$\left.\frac{dE(\varphi_t)}{dt}\right|_{t=0} = -\int_M <\tau(\varphi), v> \, dx,$$

where $\tau(\varphi) = \text{Trace}\nabla d\varphi$ is the trace of the second fundamental form of φ, called the *tension field* of φ (see Chapter I). We note that $\tau(\varphi)$ is a vector field along φ; i.e., a section of the bundle $\varphi^{-1}T(N) \to M$. Here τ represents the Euler-Lagrange operator of E, and a smooth map $\varphi : M \to N$ is harmonic if and only if $\tau(\varphi) = 0$. In local charts, that takes the form

(5) $$-\Delta\varphi^\gamma + {}^N\Gamma^\gamma_{\alpha\beta}\frac{\partial\varphi^\alpha}{\partial x_i}\frac{\partial\varphi^\beta}{\partial x_j}g^{ij} = 0 \quad (1 \leq \gamma \leq n),$$

where Δ denotes the Laplacian of the Riemannian metric g of M; and ${}^N\Gamma^\gamma_{\alpha\beta}$ the Christoffel symbols of the Levi-Civita connection of the n-dimensional manifold N. (The sign convention for Δ is such that $\Delta = -\sum_{i=1}^m \frac{\partial^2}{\partial x_i^2}$ on $\mathbb{R}^m$).

We observe that curvature in the range is responsible for the nonlinearity of (5). On the other hand, (5) is semi-linear, meaning that the system is linear in its highest order derivatives; and diagonal in the sense that the highest order operators are the same for each equation of the system. Of course, (5) is of divergence type, being the Euler-Lagrange equation of the functional E.

(6) Let $\varphi : M \to N$ be an immersion. Its volume

(7) $$V(\varphi) = \int_M |\Lambda^m d\varphi| \, dx;$$

here $|A^m d\varphi|$ denotes the Jacobian of φ; and dx is now the volume element associated with the induced metric $\varphi^* h$, where h is the metric of N.

If $(\varphi_t)_{t \in \mathbf{R}}$ is a variation of $\varphi = \varphi_0$ through immersions, we have (Chapter I (2))

$$(8) \qquad \frac{dV(\varphi_t)}{dt}\bigg|_{t=0} = -\int_M <\tau(\varphi), v> dx \ .$$

Because *minimal* immersions are, by definition, the smooth critical points of the volume functional, comparing (4) and (8) tells us that a *Riemannian* (= isometric) immersion is minimal if and only if it is harmonic. We also observe that the second fundamental form of a Riemannian immersion takes its values in the normal bundle of M in N.

(9) **Example 1.** If M is the circle S^1, then the harmonic maps $\varphi : S^1 \to N$ are just the closed geodesics of N, parametrized proportionally to arc length. In particular, in this case (5) is a system of *ordinary* differential equations

$$\frac{d^2\varphi^\gamma(t)}{dt^2} + {}^N\Gamma^\gamma_{\alpha\beta}(\varphi(t))\frac{d\varphi^\alpha(t)}{dt}\frac{d\varphi^\beta(t)}{dt} = 0 \quad (1 \leq \gamma \leq n) \ .$$

In particular, if $N = S^1$ the k-fold rotation ($k \in \mathbf{Z}$) $\theta \mapsto k\theta, 0 \leq \theta < 2\pi$, is a harmonic map of Brouwer degree k.

Example 2. If M is a closed oriented surface, then E is a conformal invariant of the metric on M. Therefore, questions of harmonicity depend only on the structure of M as a Riemann surface. Thus we have at our disposal the machinery of holomorphic function theory (e.g., isothermal charts, uniformization, holomorphic quadratic differentials, Weierstrass representations). Also,

$$(10) \qquad V(\varphi) \leq E(\varphi) \ ,$$

with equality if and only if φ is conformal (Chapter II (1.5)).

One of the consequences of (10) is this: A *non-constant map* $\varphi : M \to N$ *is a minimal branched immersion if and only if it is conformal harmonic*.

Harmonic maps $M \to S^2$ arise via the Gauss maps of Riemannian immersions of M in $\mathbf{R}^3$ with constant mean curvature (Chapters III, V). Here are some especially significant examples, due to C. Delaunay:

Roll a conic C on a line in the plane; then in $\mathbf{R}^3$ rotate about that line the trace of a focus. The resulting surface M in $\mathbf{R}^3$ has constant mean curvature. Its Gauss map is harmonic.

a) If C is a parabola, then M is a catenoid; that has mean curvature zero, and its Gauss map is anti-holomorphic.

b) If C is an ellipse, then M is Delaunay's unduloid, an embedded surface in $\mathbf{R}^3$ with non-zero constant mean curvature. Its Gauss map covers a band about the equator. Factoring out the periodicity produces a harmonic map of degree zero of a 2-torus into S^2.

c) If C is an hyperbola, then M is Delaunay's nodoid. That is an immersed surface in $\mathbf{R}^3$, again with non-zero constant mean curvature. Its Gauss map is surjective. Factoring out the periodicity produces a harmonic map of a Klein bottle $K \to S^2$ representing the non-trivial cohomology class in $H^2(K) = \mathbf{Z}_2$.

Example 3. The harmonic maps $\varphi : M \to S^n$ are the solutions of

$$\Delta \Phi = |d\Phi|^2 \Phi , \tag{11}$$

where $\Phi = i \circ \varphi$ is the composition of φ with the standard inclusion map $i : S^n \to \mathbf{R}^{n+1}$. Similarly, minimal immersions in S^n are solutions φ of (11) which are immersions, with Δ now computed with respect to the metric induced by φ. Furthermore, in this case the composition $\Phi : M \to \mathbf{R}^{n+1}$ is an immersion with parallel mean curvature.

Geometrically, (11) expresses the fact that -because i is a Riemannian immersion- $\tau(\varphi)$ vanishes identically if and only if $\tau(\Phi)(= -\Delta\Phi)$ is orthogonal to $T(S^n)$ at each point. (See I (1.17)).

If $n = 1$, the harmonic maps $\varphi : M \to S^1$ are canonically identified with the harmonic 1-forms on M with integral periods (Chapter IV(2)). Locally, they are solutions of the linear elliptic differential equation

$$g^{ij}(x)\left(\frac{\partial^2 \varphi}{\partial x_i \partial x_j}(x) - {}^M\Gamma^k_{ij}(x)\frac{\partial \varphi(x)}{\partial x_k}\right) = 0 ,$$

the left member being a representation of $\Delta \varphi$ in a local chart.

Let $M = S^1 = N$; we use complex numbers $z = x_1 + ix_2$ on $\mathbf{R}^2 = \mathbf{C}$. Let $P_k \to \mathbf{R}$ denote the polynomial $P_k(z) = \mathrm{Re}(z^k/k)$, $k \in \mathbf{Z} - \{0\}$. Then P_k is a harmonic homogeneous polynomial and its gradient $\nabla P_k|_{S^1} : S^1 \to S^1$ defines a solution of (11) with $|d\Phi|^2 = (k-1)^2$; that reproduces the harmonic maps illustrated in the last paragraph of Example 1 above.

More generally, solutions of (11) with constant energy density $\frac{1}{2}|d\Phi|^2 = \lambda/2$ are called *eigenmaps* with eigenvalue λ. Clearly their study interacts with the spectral theory of Δ to give a particularly rich source of geometric applications when $M = S^m$ (see Chapters VIII, IX). Let $f : \mathbf{R}^2 \times \mathbf{R}^2 \to \mathbf{R}^2$ denote complex multiplication. We define

$$H : \mathbf{R}^4 \to \mathbf{R}^3 \quad \text{by} \quad (z_1, z_2) \mapsto \left(2f(z_1, z_2), |z_1|^2 - |z_2|^2\right) .$$

Then $h = H|_{S^3} : S^3 \to S^2$ is an *eigenmap*; i.e., a harmonic map with constant energy density equal to 4. The map h is called the *Hopf fibration* (see Chapter VIII (2.8)).

Example 4. Let M and P be Riemannian manifolds and $\pi : M \to P$ a submersion (i.e., each differential $d\pi(x) : T_x(M) \to T_{\pi(x)}(P)$ is surjective); then there is a canonical splitting of the tangent bundle

$$T(M) = T^H(M) \oplus T^V(M)$$

where $T^V(M) = Ker\, d\pi$ and $T^H(M)$ is its orthogonal complement. We say that π is a *Riemannian submersion* if each restriction $d\pi(x) : T_x^H(M) \to T_{\pi(x)}(P)$ is an isometry. We have the following basic characterization: *A Riemannian submersion $\pi : M \to P$ is harmonic if and only if its fibres are minimal submanifolds.* The Hopf fibration $h : S^3 \to S^2(1/2)$ is one such, the fibres being geodesic circles on S^3. Both eigenmaps and Riemannian submersions will play a fundamental role in our reduction theory of Chapter IV and its applications.

(12) Global existence of solutions of (5) cannot be expected, in general - although there is a satisfactory theory when the sectional curvature of N is non-positive. In this case we have existence of E-minima; however, the methods required are entirely different from those developed in this monograph. In geometric terms, if M and N are compact, we look for solutions $\varphi : M \to N$ of (5) belonging to a prescribed homotopy class. We cannot always find one: there is no harmonic map of degree one from a 2-torus to the 2-sphere. (We shall restrict attention to manifolds without boundary - we do so reluctantly, for that excludes many fine applications). Altogether, we can expect severe difficulties in establishing existence if the range is positively curved. More specially, in view of the observations i), ii) in (3) above, most solutions will be unstable; in particular, not even local minima of E; see also Appendix 1.

(13) Minimal immersions (with or without boundary) have important applications in Riemannian geometry. However, we have no systematic approach to their construction; most of the known examples arise in some sort of symmetric context. For instance, let G be a compact Lie group of isometries of N. Its action provides a singular foliation of N by orbits $\Gamma = G/H$, whose types depend on the isotropy subgroups H. (See Chapter IV and V for a detailed exposition of this and the next paragraphs). The volume function $V : N/G \to \mathbb{R}$ on the orbit space associates to an orbit Γ its volume $V(\Gamma)$ in N; then a critical point of that function is a homogeneous minimal immersion. That led W-Y. Hsiang to the conclusion that *every compact homogeneous Riemannian manifold G/H can be minimally immersed in some Euclidean sphere* (See Chapter V (1.13)).

(14) In the presence of suitable equivariance system (11) reduces to an ordinary differential equation, possibly with singular boundary values (Chapter VI) or prescribed asymptotic limits (Chapters IX, X). One of our goals is to illustrate some favourable cases where the analysis can be carried through to establish existence of non-trivial minimal immersions or harmonic maps; indeed, we are as interested in the methods of solution as in their geometrical or topological properties. It is worth noting that in most cases the relevant ordinary differential equations depend on parameters intimately related to the geometry of the problem ; and these restrict the existence and qualitative behaviour of solutions.

(15) **Example.** (From Chapter VI). In the notation of (13) above, we take $N = S^4$, $G = SO(2) \times SO(2)$. Then the space $Q = N/G$ can be identified with a spherical sector in S^2:

$$Q = \{(\eta, \theta) : 0 \leq \eta \leq \pi, 0 \leq \theta \leq \pi/2\}$$

with induced metric $h = \sin^2 \eta \, d\theta^2 + d\eta^2$ and volume function $V = V(\eta, \theta) = \sin^2 \eta \sin 2\theta$.

(16) Let $\gamma(s) = (\eta(s), \theta(s))$ be a curve in Q which starts at a point $(\eta_0, 0)$ and ends at $(\eta_1, \pi/2)$ without self-intersections, $0 < \eta_0, \eta_1 < \pi$. Then $\sigma^{-1}(\gamma) \subset S^4$ is an embedded 3-sphere ($\sigma : S^4 \to Q$ denoting the canonical projection). Our reduction theory shows that $\sigma^{-1}(\gamma)$ is *minimal in S^4 if and only if*

(17) $$\begin{cases} \ddot{\eta} - 3\cos\eta \sin\eta \, \dot{\theta}^2 + 2\cot(2\theta)\dot{\theta}\dot{\eta} = 0 \\ \dot{\eta}^2 + \dot{\theta}^2 \sin^2\eta = 1. \end{cases}$$

That ordinary differential system (17) admits special solutions as in (16) - a fact which proves the existence of non-equatorial minimal embeddings of S^3 into S^4. A variation of this construction can be used to obtain non-standard immersions of S^n in $\mathbf{R}^{n+1}$ with constant mean curvature, for all $n \geq 3$.

(18) We remark that in general - as the previous example shows - the orbit space N/G is not a manifold. The geometric singularity of N/G has its analytical counterpart in the singularity of the coefficients of the associated differential equation - see (17), which is singular at $\theta = 0$ and $\theta = \pi/2$; and also (21) below, at $s = 0$ and $s = \pi/2$.
That fact is responsible for many of the analytical difficulties encountered in our study of the qualitative behaviour of solutions.

(19) **Example.** (From Chapter IX). We can write the Euclidean $(p+r-1)$-sphere S^{p+r-1} as the join $S^{p-1} * S^{r-1}$, whose points are parametrized by $(\sin s.x, \cos s.y)$ with $x \in S^{p-1}$, $y \in S^{r-1}$ and $0 \leq s \leq \pi/2$. A map $\varphi : S^{p-1} \to S^{q-1}$ is

an eigenmap if it satisfies (11) with constant eigenvalue $\lambda_\varphi = |d\Phi|^2$. Now take two eigenmaps $u : S^{p-1} \to S^{q-1}$ and $v : S^{r-1} \to S^{s-1}$, and form their join $u * v : S^{p+r-1} \to S^{q+s-1}$. If $\alpha : [0, \pi/2] \to [0, \pi/2]$ is a function satisfying the boundary conditions

(20) $\qquad\qquad\qquad \alpha(0) = 0 \quad$ and $\quad \alpha(\pi/2) = \pi/2$,

the α-join

$$u *_\alpha v : S^{p+r-1} \to S^{q+s-1}$$

is given by

$$(\sin s.x, \cos s.y) \mapsto (\sin \alpha(s).u(x), \cos \alpha(s).v(y)).$$

For simplicity we take $p, r > 2$. Then (up to a constant factor) the energy $E(u *_\alpha v)$ reduces to the 1-dimensional integral

$$J(\alpha) = \int_0^{\pi/2} \left[\dot\alpha^2(s) + \frac{\sin^2 \alpha(s)}{\sin^2 s}\lambda_u + \frac{\cos^2 \alpha(s)}{\cos^2 s}\lambda_v\right] \sin^{p-1} s \cos^{r-1} s \, ds.$$

The Euler-Lagrange equation associated to J is

(21) $\qquad\qquad\qquad \ddot\alpha(s) + D(s)\dot\alpha(s) - G(s, \alpha(s)) = 0$,

where

$$D(s) = (p-1)\cot s - (r-1)\tan s$$

and

$$G(s, \alpha(s)) = \left(\frac{\lambda_u}{\sin^2 s} - \frac{\lambda_v}{\cos^2 s}\right) \sin \alpha(s) \cos \alpha(s).$$

We can view (21) as a description of the motion of a pendulum, damped according to the function D and with variable gravity G.

A solution α of (21) with boundary values (20) determines a harmonic map $u *_\alpha v$ homotopic to $u * v$. For instance, we can apply the direct method of variational theory to J to obtain a harmonic join $u *_\alpha v$ if

$$(r-2)^2 < 4\lambda_v \quad \text{and} \quad (p-2)^2 < 4\lambda_u.$$

Again, we emphasize that the harmonic map $u *_\alpha v$ so obtained will often be an unstable critical point of E - even though the function α is an absolute minimum of J.

(22)   Our problems belong in particular to the differential geometry of fibre bundles; elliptic variational theory - with special emphasis on unstable critical points; ordinary differential equations, often of non-standard varieties.

However, our primary objective is to present representative examples to illustrate the reduction methods and associated analysis - with geometric and topological applications. In most existence theorems we do not know whether the results obtained are best possible; or are restricted only because of the methods used. For instance,
(i) we prove W-Y. Hsiang's theorem that there are non-equatorial minimally embedded hyperspheres in S^n for $n = 4$. Variations on the method give the same conclusion for many other dimensions (see VI (4.4)); but not all, as yet.
(ii) We use a version of R.T. Smith's method to show that every homotopy class H of maps $S^n \to S^n$ can be harmonically represented, provided that the range is given a suitable ellipsoidal metric (depending only on n). If $n \leq 7$ or $n = 9$, then we obtain the same conclusion for Euclidean spheres; but not for other values of n, as yet.

(23)   The monograph is in three parts, subdivided into ten chapters and four appendices. Each chapter has a brief introduction pointing the way and providing basic motivation. We draw special attention to those now, for they provide a perspective on the main results presented in the text. Each chapter ends with a section on notes and comments; these give indications of further or alternative directions. We have tried to produce a unified and reasonably self-contained treatment of the many different analytical/geometrical ideas which arise. For further background, applications, and origins of the key ideas, we refer to [EL1, 2], as well as [EL 3] and [L3].

(24) **Conventions.** We shall maintain the notation of [EL1, 2]. All manifolds, maps, metrics and bundles are smooth ($= C^\infty$) unless specifically stated otherwise. If $\varphi : M \to N$ is a map between Riemannian manifolds, we sometimes write $\varphi : (M, g) \to (N, h)$ to draw attention to the specific metrics under consideration. We set $m = \dim M$, $n = \dim N$. If $\pi : V \to M$ is a vector bundle, then $C(V)$ denotes the vector space of its smooth sections. ∇ will always signify the covariant derivative of sections in a vector bundle - with a superscript if necessary; thus ∇^M is the standard Levi-Civita connection on the tangent bundle $T(M)$ of M. $S^n(r)$ denotes a Euclidean sphere in $\mathbb{R}^{n+1}$ of radius r; $S^n = S^n(1)$. The reference (3.7) is to be found in the same chapter. V(3.7) is in Chapter V.

(25) **Acknowledgements.** We have benefited from suggestions by many friends. We are especially indebted to John Rawnsley for permission to use his manuscript [Ra], and TEXnical assistance; and to Helen Robinson for her comments. This

monograph was written while the second named author was a Research Fellow of the Science and Engineering Research Council. We express our gratitude to all.

April 1992

Table of Contents

INTRODUCTION

PART I Basic Variational and Geometrical Properties

Chapter I *Harmonic maps and minimal immersions*

Introduction	13
Basic properties of harmonic maps	13
Minimal immersions	20
Notes and comments	22

Chapter II *Immersions of parallel mean curvature*

Introduction	24
Parallel mean curvature	24
Alexandrov's theorem	29
Notes and comments	32

Chapter III *Surfaces of parallel mean curvature*

Introduction	34
Theorems of Chern and Ruh–Vilms	34
Theorems of Almgren–Calabi and Hopf	37
On the Sinh–Gordon equation	40
Wente's theorem	42
Notes and comments	45

Chapter IV *Reduction techniques*

Introduction	47
Riemannian submersions	48
Harmonic morphisms and maps into a circle	51
Isoparametric maps	54
Reduction techniques	58
Notes and comments	63

PART II *G*–Invariant Minimal and Constant Mean Curvature Immersions

Chapter V *First examples of reductions*

Introduction	64
G–equivariant harmonic maps	64

Rotation hypersurfaces in spheres	74
Constant mean curvature rotation hypersurfaces in $\mathbb{R}^n$	81
Notes and comments	86

Chapter VI *Minimal embeddings of hyperspheres in S^4*

Introduction	92
Derivation of the equation and main theorem	92
Existence of solutions starting at the boundary	95
Analysis of the O.D.E. and proof of the main theorem	102
Notes and comments	110

Chapter VII *Constant mean curvature immersions of hyperspheres into $\mathbb{R}^n$*

Introduction	111
Statement of the main theorem	111
Analytical lemmas	114
Proof of the main theorem	120
Notes and comments	127

PART III   Harmonic Maps Between Spheres

Chapter VIII *Polynomial maps*

Introduction	128
Eigenmaps $S^m \to S^n$	129
Orthogonal multiplications and related constructions	137
Polynomial maps between spheres	143
Notes and comments	148

Chapter IX *Existence of harmonic joins*

Introduction	151
The reduction equation	151
Properties of the reduced energy functional J	154
Analysis of the O.D.E.	157
The damping conditions	161
Examples of harmonic maps	167
Notes and comments	169

Chapter X *The harmonic Hopf construction*

Introduction	171
The existence theorem	171

Examples of harmonic Hopf constructions	179
$\pi_3(S^2)$ and harmonic morphisms	182
Notes and comments	186
Appendix 1 *Second variations*	188
Appendix 2 *Riemannian immersions $S^m \to S^n$*	200
Appendix 3 *Minimal graphs and pendent drops*	204
Appendix 4 *Further aspects of pendulum type equations*	208
References	213
Index	224

PART 1. BASIC VARIATIONAL AND GEOMETRICAL PROPERTIES

CHAPTER I. HARMONIC MAPS AND MINIMAL IMMERSIONS

INTRODUCTION

In this Chapter we describe briefly some variational theory

(1) of the energy functional E of maps $\varphi : M \to N$ between Riemannian manifolds:

$$E(\varphi) = \frac{1}{2} \int_M |d\varphi|^2 \, dx$$

(2) of the volume functional V of Riemannian (= isometric) immersions $\varphi : M \to N$:

$$V(\varphi) = \int_M |\Lambda^m \, d\varphi| \, dx \, .$$

In both cases we derive their first variational formulas. Although these are very different variational problems, there is a fundamental relationship between their smooth critical points, expressed in Corollary (2.11): *A Riemannian immersion $\varphi : M \to N$ is minimal (i.e., a critical point of V) iff φ is harmonic (i.e., a critical point of E).*

1. BASIC PROPERTIES OF HARMONIC MAPS

(1.1)   If $\varphi : M \to N$ is a smooth map between manifolds of dimensions m, n respectively, we let

$$\varphi^{-1}T(N) = \{(x, v) \in M \times T(N) : \varphi(x) = \pi(v)\} \, .$$

That is the vector bundle over M induced by φ from the tangent vector bundle $\pi : T(N) \to N$ of N; we have the natural commutative diagram

$$\begin{array}{ccc} \varphi^{-1}T(N) & \longrightarrow & T(N) \\ \downarrow & & \downarrow \pi \\ M & \longrightarrow & N \\ & \varphi & \end{array}$$

The sections of the bundle induced by φ are called *(infinitesimal) variations of φ*; or *vector fields along φ*.

At each point $x \in M$ the differential

$$d\varphi(x) : T_x(M) \to T_{\varphi(x)}(N)$$

is a linear map; thus it can be interpreted as an element of the vector space $T_x^*(M) \otimes T_{\varphi(x)}(N)$. We thereby obtain a section $d\varphi$ of the bundle

$$T^*(M) \otimes \varphi^{-1}T(N) \to M; \quad \text{i.e.,} \quad d\varphi \in C(T^*(M) \otimes \varphi^{-1}T(N)).$$

(1.2) Now suppose that g, h are Riemannian metrics on M, N respectively. The *first fundamental form of* φ is the symmetric semidefinite 2–covariant tensor field φ^*h on M. In charts,

$$(\varphi^*h)_{ij}(x) = \varphi_i^\alpha(x)\varphi_j^\beta(x)h_{\alpha\beta}(\varphi(x)), \quad \text{where} \quad \varphi_i^\alpha = \frac{\partial\varphi^\alpha}{\partial x^i}.$$

(1.3) We define the *energy density of* φ:

$$e(\varphi) = \frac{1}{2}\,\text{Trace}_g(\varphi^*h) : M \to \mathbb{R}(\geq 0); \quad \text{i.e.,}$$

$$e(\varphi)(x) = \frac{1}{2}g^{ij}(x)\varphi_i^\alpha(x)\varphi_j^\beta(x)h_{\alpha\beta}(\varphi(x)) = \frac{1}{2}|d\varphi(x)|^2$$

where the vertical bars refer to the Hilbert–Schmidt norm on

$$T_x^*(M) \otimes T_{\varphi(x)}(N) = \text{Hom}(T_x(M), T_{\varphi(x)}(N)).$$

We can also write

$$|d\varphi(x)|^2 = \sum_i |d\varphi(x)e_i|^2_{\varphi(x)}$$

in terms of an orthonormal base $(e_i)_{1 \leq i \leq m}$ on $T_x(M)$, where $|\ |_{\varphi(x)}$ refers to the norm of $T_{\varphi(x)}(N)$.

Observe that (by taking normal charts) $e(\varphi) \equiv 0$ on an open set iff φ is constant there.

(1.4) On a compact domain $M' \subset M$ the energy of φ is

$$E(\varphi; M') = \int_{M'} e(\varphi)(x)dx,$$

where $dx = \sqrt{\det g(x)}\,dx^1\ldots dx^m$ is the canonical measure associated with the metric g. Our primary interest is in the case where M is compact; we write $E(\varphi) = E(\varphi; M)$.

(1.5) The Levi–Civita connections of g and h induce connections on $T^*(M)$ and $\varphi^{-1}T(N)$; and thereby on their tensor product:

$$\nabla = \nabla^{T^*(M)} \otimes \nabla^\varphi .$$

The second fundamental form of a map φ is the covariant differential of $d\varphi$:

$$\nabla d\varphi \in \mathcal{C}(\odot^2 T^*(M) \otimes \varphi^{-1}T(N)) .$$

To verify that $\nabla d\varphi$ is indeed symmetric, we compute

$$\nabla d\varphi(X,Y) = (\nabla_X d\varphi)Y = \nabla_X^\varphi(d\varphi \cdot Y) - d\varphi(\nabla_X^M Y)$$

for $X, Y \in \mathcal{C}(T(M))$; and then

$$\nabla d\varphi(X,Y) - \nabla d\varphi(Y,X) = \nabla_X^\varphi(d\varphi \cdot Y) - \nabla_Y^\varphi(d\varphi \cdot X) - d\varphi(\nabla_X^M Y - \nabla_Y^M X)$$
$$= d\varphi \cdot [X,Y] - d\varphi \cdot [X,Y] = 0. \ //$$

In charts

$$(\nabla d\varphi)_{ij}^\gamma = (\nabla d\varphi^\gamma)_{ij} + {}^N\Gamma_{\alpha\beta}^\gamma \varphi_i^\alpha \varphi_j^\beta \quad (1 \le i, j \le m, \ 1 \le \gamma \le n)$$

where ${}^N\Gamma_{\alpha\beta}^\gamma$ are the Christoffel symbols of (N, h).

We shall use the notations $\nabla d\varphi = \beta(\varphi) = \beta_\varphi$ indifferently.

(1.6) Say that a map $\varphi : M \to N$ *is totally geodesic if* $\nabla d\varphi \equiv 0$. It is easy to see that that property is characterized by the condition that φ carries geodesics of M linearly to geodesics of N.

A totally geodesic map has constant energy density $m/2$ *and constant rank*; indeed, set $r = \max\{\text{rank } d\varphi(x) : x \in M\}$. Then a computation shows

$$d|\Lambda^r d\varphi|^2 = 2 < \nabla \Lambda^r d\varphi, \Lambda^r d\varphi > = 0 ,$$

so $|\Lambda^r d\varphi|$ is constant on M.

(1.7) The *tension field* $\tau(\varphi)$ is defined by

$$\tau(\varphi) = \text{Trace}_g \nabla d\varphi .$$

Thus $\tau(\varphi)$ is a vector field along φ:

$$\tau(\varphi) \in \mathcal{C}(\varphi^{-1}T(N)) .$$

In charts
$$\tau(\varphi) = g^{ij}(\nabla_{\partial_i} d\varphi)(\partial_j); \quad \text{i.e.,}$$
$$\tau^\gamma(\varphi) = g^{ij}(\nabla d\varphi^\gamma)_{ij} + g^{ij}\,{}^N\Gamma^\gamma_{\alpha\beta}\,\varphi_i^\alpha\,\varphi_j^\beta$$
$$= -\Delta\varphi^\gamma + g^{ij}\,{}^N\Gamma^\gamma_{\alpha\beta}\,\varphi_i^\alpha\,\varphi_j^\beta \quad (1 \leq \gamma \leq n),$$

where Δ is the Laplace operator on M. We use the sign convention $\Delta = -\text{div}\,\nabla$ on functions.

A map $\varphi : M \to N$ is *harmonic* iff $\tau(\varphi) \equiv 0$.

The system of partial differential equations $\tau(\varphi) \equiv 0$ is

(1) semilinear elliptic (*semilinear* means that the system is linear in its highest order derivatives; elliptic because in each equation those highest order terms coincide with those of the Laplacian Δ of M);

(2) in divergence form;

(3) in diagonal form (the highest order operators in the equations coincide);

(4) quadratic in the first derivatives.

(1.8) A map $\varphi : M \to N$ is *a critical point* (or *an extremum*) *of E* if

$$\left.\frac{dE(\varphi_t)}{dt}\right|_{t=0} = 0 \quad \text{for all deformations } (\varphi_t)_{t\in\mathbf{R}}$$

of $\varphi = \varphi_0$. For that it is sufficient to examine those deformations of the form

$$\varphi_t(x) = \exp_{\varphi(x)} tv(x) \quad \text{for variations } v \in \mathcal{C}(\varphi^{-1}T(N)).$$

(1.9) **Theorem.** *A map φ is a critical point of E iff it is harmonic.*

Proof. For simplicity we shall assume that M is compact. (Otherwise, work in compact domains, using compactly supported variations.)

Now define $\Phi : M \times \mathbf{R} \to N$ by $\Phi(x, t) = \varphi_t(x)$. Then $d\varphi_t$ is the differential along M for fixed t; and ∇_{∂_t} the covariant derivative in $T^*(M \times \mathbf{R}) \otimes \Phi^{-1}T(N)$.

Step 1. For any deformation $(\varphi_t)_{t\in\mathbf{R}}$

$$\left.\frac{d}{dt} E(\varphi_t)\right|_{t=0} = \frac{1}{2}\int_M \frac{\partial}{\partial t} <d\varphi_t, d\varphi_t>\Big|_0 dx$$
$$= \int_M <\nabla_{\partial_t} d\varphi_t, d\varphi_t>\Big|_0 dx.$$

But for any vector field $X \in C(T(M))$,

$$(\nabla_{\partial_t} d\varphi_t)(X) = \nabla^{\Phi}_{\partial_t}(d\varphi_t \cdot X) - d\varphi_t \cdot \nabla^{T(M \times \mathbb{R})}_{\partial_t} X$$
$$= \nabla^{\Phi}_X (d\Phi \cdot \partial_t) + 0 .$$

Therefore

$$\left.\frac{dE(\varphi_t)}{dt}\right|_0 = \int_M <\nabla^{\Phi} \frac{\partial \Phi}{\partial t}, d\varphi_t>\bigg|_0 dx = \int_M <\nabla^{\varphi} v, d\varphi> dx ,$$

where $v = \left.\frac{\partial \varphi_t}{\partial t}\right|_0 \in C(\varphi^{-1}T(N))$.

Step 2. Write

$$\int_M <\nabla^{\varphi} v, d\varphi> dx = \int_M <v, \nabla^* d\varphi> dx .$$

For $\rho \in C(T^*(M) \otimes \varphi^{-1}T(N))$ we have

$$\nabla^* \rho = -g^{su}(\nabla_{\partial_s} \rho)(\partial_u) .$$

Indeed, take any point $x_0 \in M$ and an x_0-centred chart with coordinate vectors $(\partial_i)_{1 \leq i \leq m}$ such that

$$\nabla_{\partial_i} \partial_j = 0, \quad \nabla_{\partial_i} g^{su} = 0 \quad \text{at} \quad x_0 .$$

Then for any variation $v \in C(\varphi^{-1}T(N))$,

$$<\nabla \sigma, \rho> - <\sigma, -g^{su}(\nabla_{\partial_s} \rho)(\partial_u) >$$
$$= <\nabla_{\partial_s} \sigma, \rho(\partial_u)> g^{su} + <\sigma, (\nabla_{\partial_s} \rho)(\partial_u)> g^{su}$$
$$= \nabla_{\partial_s} <\sigma, \rho(\partial_u)> g^{su} ,$$

which is the divergence of a vector field. Now we apply Green's theorem to obtain

$$(1.10) \qquad \int_M <\nabla \sigma, \rho> dx = \int_M <\sigma, -g^{su}(\nabla_{\partial_s} \rho)(\partial_u)> dx .$$

Because that is true for all σ, we conclude that

$$\nabla^* \rho = -g^{su}(\nabla_{\partial_s} \rho)(\partial_u) .$$

Step 3. Taking $\rho = d\varphi$ gives

$$(1.11) \qquad \left.\frac{dE(\varphi_t)}{dt}\right|_0 = \int_M <\nabla^{\varphi} v, d\varphi> dx = -\int_M <v, \tau(\varphi)> dx .$$

That is valid for all deformations (φ_t) iff $\tau(\varphi) \equiv 0$. //

$\tau(\varphi)$ is the *Euler–Lagrange operator associated to the energy functional* E.

(1.12) **Example.** Take $M = S^1$, the unit circle. Then
$$E(\varphi) = \frac{1}{2} \int_{S^1} |\varphi'(s)|^2 ds ,$$
with Euler–Lagrange operator $\tau(\varphi) = \nabla_{\varphi'} \varphi'$, which is the acceleration vector field of φ. Its critical points are the closed geodesics of N, parametrized proportionally to arc length.

(1.13) **Example.** A function $\varphi : M \to \mathbb{R}$ is a harmonic map iff
$$-\Delta\varphi = g^{ij}(\nabla d\varphi)_{ij} = g^{ij}\left(\frac{\partial^2 \varphi}{\partial x^i \partial x^j} - {}^M\Gamma_{ij}^k \frac{\partial \varphi}{\partial x^k}\right) \equiv 0 .$$
I.e., iff φ is a harmonic function.

If M is compact, then φ is harmonic iff it is constant. Indeed, taking $\rho = d\varphi \in C(T^*(M))$ in (1.10), we have
$$\int_M <\nabla\sigma, \rho> dx = \int_M <\sigma, \nabla^*\rho> dx = \int_M <\sigma, \Delta\varphi> dx .$$
Setting $\sigma = \varphi$ gives
$$\int_M |d\varphi|^2 dx = \int_M <\varphi, \Delta\varphi> dx ,$$
from which the assertion follows.

That is also a consequence of the maximum principle ((1.18) below).

(1.14) **Proposition.** *Given maps* $M \xrightarrow{\varphi} N \xrightarrow{\psi} P$, *we have* $\nabla d(\psi \circ \varphi) = d\psi \cdot \nabla d\varphi + \nabla d\psi(d\varphi, d\varphi)$; *and* $\tau(\psi \circ \varphi) = d\psi \cdot \tau(\varphi) + \text{Trace } \nabla d\psi(d\varphi, d\varphi)$.

Proof.
$$\nabla d(\psi \circ \varphi)(X, Y) = \nabla_X(d\psi \circ d\varphi \cdot Y) - d(\psi \circ \varphi) \cdot \nabla_X Y$$
$$= (\nabla_{d\varphi \cdot X} d\psi) d\varphi \cdot Y + d\psi \cdot \nabla_X(d\varphi \cdot Y) - d\psi \circ d\varphi \cdot \nabla_X Y$$
$$= \nabla d\psi(d\varphi \cdot X, d\varphi \cdot Y) + d\psi \cdot \nabla d\varphi(X, Y) .//$$

(1.15) Corollary. *If φ is harmonic and ψ totally geodesic, then $\psi \circ \varphi$ is harmonic.*

Say that $\psi : N \to P$ is a *Riemannian immersion* if $h = \psi^* k$, where h, k are the metrics of N, P, respectively.

(1.16) Corollary. *Let $\psi : N \to P$ be a Riemannian immersion. Set*

$$\Phi = \psi \circ \varphi : M \to P.$$

Then $d\psi \cdot \tau(\varphi)$ is the tangential component, and Trace $\nabla d\psi(d\varphi, d\varphi)$ the normal component of $\tau(\Phi)$. In particular, φ is harmonic iff $\tau(\Phi) \perp N$.

(1.17) Proposition. *Let $i : S^n \to \mathbb{R}^{n+1}$ be the standard inclusion. A map $\varphi : M \to S^n$ is harmonic iff*

$$\Delta \Phi = |d\Phi|^2 \Phi,$$

where $\Phi = i \circ \varphi$. And $|d\varphi|^2 = |d\Phi|^2$.

Proof. Because i is a Riemannian immersion and $\tau(\Phi) = -\Delta\Phi$, Corollary (1.16) tells us that φ is harmonic iff $\Delta\Phi = A\Phi$ for some function $A : M \to \mathbb{R}$. But $|\Phi(x)|^2 \equiv 1$, so $0 = \frac{1}{2}\Delta|\Phi(x)|^2 = <\Phi(x), \Delta\Phi(x)> - |d\Phi(x)|^2$. Thus $A = |d\Phi|^2$. Moreover, $|d\varphi|^2 = |d\Phi|^2$ because i is a Riemannian immersion. //

We will need two fundamental results of elliptic analysis; the first is the maximum principle, which we state in a form suitable for application to the proof of Alexandrov's Theorem (II (2.6)).

(1.18) Maximum principle. *Let U be a connected open subset of $\mathbb{R}^m$ and $\varphi : U \to \mathbb{R}$ a C^2-solution of the linear elliptic differential equation*

$$(1.19) \qquad \sum_{i,j=1}^{m} A_{ij} \frac{\partial^2 \varphi}{\partial x^i \partial x^j} + \sum_{k=1}^{m} B_k \frac{\partial \varphi}{\partial x^k} = 0$$

whose coefficients $A_{ij} = A_{ji}$ and B_k are continuous.

Case 1. *If φ has a non–negative relative maximum at some point $p \in U$, then φ is constant.*

Case 2. *Assume that φ has a non–negative relative maximum on $U \cup \{p\}$ for some point $p \in \partial U$ at which the coefficients are continuous and (1.19) is still elliptic. And assume that there is a closed ball $\overline{B_\rho(q)}$ such that*

(i) $\overline{B_\rho(q)} \subset U \cup \{p\}$

(ii) $p \in \partial B_\rho(q)$

(iii) *the directional derivative of φ at p in the direction from p to q is 0.*

Then φ is constant.

(1.20) Our main occupation is with *smooth* maps of Riemannian manifolds. However, the energy functional $E(\varphi)$ is naturally defined on the Sobolev space $\mathcal{L}_1^2(M, N)$. In general, extremals of E in $\mathcal{L}_1^2(M, N)$ (i.e., the weakly harmonic maps) are not continuous, so they only satisfy the tension field equations weakly. However, we have the

(1.21) **Regularity Theorem.** *Any continuous weakly harmonic map is harmonic (and therefore smooth).*

2. MINIMAL IMMERSIONS

(2.1) If $\varphi : M \to N$ is a Riemannian immersion, at each point $x \in M$ the Jacobian

$$J_\varphi(x) = \Lambda^m d\varphi(x) : \Lambda^m T_x(M) \to \Lambda^m T_{\varphi(x)}(N)$$

is injective. Identifying a vector field $X \in \mathcal{C}(T(M))$ with $d\varphi \cdot X \in \mathcal{C}(\varphi^{-1}T(N))$, we obtain

(2.2) $$\nabla_X^N Y = \nabla_X^M Y + \nabla d\varphi(X, Y).$$

That describes the decomposition of the connection of N (restricted to M) into its tangential component (i.e., the connection on M) and its normal component. The latter is the second fundamental form of φ in the classical sense ([KN II (Ch. VII)]) used in the theory of Riemannian immersions.

(2.3) Traditionally a Riemannian immersion $\varphi : M \to (N, h)$ is said to be *minimal* if it is an extremum of the volume functional

(2.4) $$V(\varphi) = \int_M |\Lambda^m d\varphi(x)| dx$$

with respect to variations (φ_t) through Riemannian immersions. We take the induced metric $\varphi^* h$ on M; thus $|\Lambda^m d\varphi(x)| dx$ is the associated volume form.

Let $I = [-1, +1]$ and $(\varphi_t)_{t \in I}$ be a deformation of $\varphi = \varphi_0$. Supposing M compact for simplicity of exposition, we have

(2.5) Theorem.
$$\left.\frac{dV(\varphi_t)}{dt}\right|_0 = -\int_M \left< \tau(\varphi), \left.\frac{\partial \varphi_t}{\partial t}\right|_0 \right> \nu$$

where ν denotes the volume element associated with the induced metric $\varphi_0^* h$.

Proof. Step 1. The volume form of the induced metric $g(t) = \varphi_t^* h$ is $\nu(t) = \sqrt{\det g(t)}\, \nu$, so that

(2.6) $$V(\varphi_t) = \int_M \nu(t).$$

(2.7) Lemma. $\left.\dfrac{\partial \det g(t)}{\partial t}\right|_0 = \text{Trace } \left.\dfrac{\partial g(t)}{\partial t}\right|_0.$

Proof. That is a property of 1-parameter families of endomorphisms $(g(t))_{t \in I}$ of a Euclidean space, with $g(0) = Id$. Let $(e_i)_{1 \le i \le m}$ be an orthonormal base at a point (with respect to $g(0)$), and let $\nu = e^1 \wedge \ldots \wedge e^m$ the associated volume form, where (e^i) is the dual base of (e_i); in particular,

$$\nu(e_1, \ldots, e_m) = 1.$$

Then

$$\frac{\partial}{\partial t} \det g(t) = \frac{\partial}{\partial t} \nu(g(t)e_1, \ldots, g(t)e_m) = \sum_{k=1}^m \nu(g(t)e_1, \ldots, \frac{\partial g(t)}{\partial t} e_k, \ldots, g(t)e_m).$$

Setting $t = 0$ in this expression gives $\sum_{k=1}^m \dfrac{\partial g_{kk}(0)}{\partial t}$, which is just Trace $\dfrac{\partial g(0)}{\partial t}$. //

Then

$$\frac{\partial \nu(t)}{\partial t} = \frac{1}{2} [\det g(t)]^{-1/2} \frac{\partial \det g(t)}{\partial t} \nu,$$

so

(2.8) $$\left.\frac{\partial \nu(t)}{\partial t}\right|_0 = \frac{1}{2} \text{Trace } \left.\frac{\partial g(t)}{\partial t}\right|_0 \nu$$

by Lemma (2.7).

Step 2. Take an orthonormal frame field $(e_i)_{1 \le i \le m}$ in a neighbourhood U_x in M, centred at x and satisfying $(\nabla_{e_i} e_j)_x = 0$ for all $1 \le i, j \le m$. Adjoin $e_0 = \partial/\partial t$, and extend the frame field $(e_i)_{0 \le i \le m}$ over $I \times U_x$. Then $[e_0, e_k] = 0$, so $\nabla_{e_0} e_i =$

$\nabla_{e_i} e_0$. Let $(\tilde{e}_k)_{0 \leq k \leq m}$ be their images under the map φ_t. Note that $(\nabla_{e_i} e_j)_x$ is the tangential component of $(\nabla_{\tilde{e}_i} \tilde{e}_j)_{(0,x)}$ in $T_{\varphi(x)}(N)$. Then

(2.9)
$$\frac{\partial g(t)}{\partial t}(\tilde{e}_i, \tilde{e}_j) = <\nabla_{\tilde{e}_0}\tilde{e}_i, \tilde{e}_j> + <\tilde{e}_i, \nabla_{\tilde{e}_0}\tilde{e}_j> = <\nabla_{\tilde{e}_i}\tilde{e}_0, \tilde{e}_j> + <\tilde{e}_i, \nabla_{\tilde{e}_j}\tilde{e}_0>$$

Step 3. Note that by definition, at $t = 0$ we have

(2.10)
$$\tilde{e}_0(0, x) = \tilde{e}_0 = \varphi_* \frac{\partial}{\partial t} = \frac{\partial \varphi_t}{\partial t}\bigg|_{t=0}.$$

If $\tilde{e}_0$ is everywhere normal to M, then

$$\frac{\partial g(t)}{\partial t}\bigg|_0 = -2 <\nabla d\varphi, \tilde{e}_0>.$$

Indeed, by (2.9)

$$\frac{\partial g(t)}{\partial t}(\tilde{e}_i, \tilde{e}_j)|_0 = <\nabla_{\tilde{e}_i}\tilde{e}_0, \tilde{e}_j> + <\tilde{e}_i, \nabla_{\tilde{e}_j}\tilde{e}_0>$$
$$= -2 <\nabla_{\tilde{e}_i}\tilde{e}_j, \tilde{e}_0> = -2 <(\nabla d\varphi)_{ij}, \tilde{e}_0>.$$

Therefore by (2.8)
$$\frac{\partial \nu(0)}{\partial t} = - <\tau(\varphi), \tilde{e}_0> \nu.$$

That, together with (2.6) and (2.10), complete the proof of Theorem (2.5). //

(2.11) **Corollary.** *A Riemannian immersion $\varphi : M \to N$ is minimal iff φ is harmonic.*

3. Notes and comments

(3.1) The basic notions relating to harmonic maps can be developed systematically in terms of the calculus of connections on vector bundles. A full account is given in [EL3].

(3.2) A delightful reference for the elementary theory of minimal submanifolds is [L3].

(3.3) Corollary (2.11) appeared in [ES].

(3.4) The maximum principle (1.18) is due to Hopf [Ho2].

(3.5) The regularity theorem (1.21) is based on fundamental contributions by Morrey [Mo] and Ladyzenskaya-Ural'ceva[LU], with present form due to Hildebrandt. Schoen and Uhlenbeck [SU1,2] have developed a partial regularity theory for energy-minimizing maps.

(3.6) Let $\varphi : M \to S^n$ be a minimal immersion of a compact manifold. Set $r = 2 - 1/(n-m)$. If for every point $x \in M$

$$|\beta_\varphi(x)|^2 \leq m/r,$$

with strict inequality somewhere, then φ is totally geodesic. If $|\beta_\varphi|^2 \equiv m/r$, then M is either
(a) $S^p\left(\sqrt{\frac{p}{m}}\right) \times S^q\left(\sqrt{\frac{q}{m}}\right)$ for some $0 < p < m$ and $p+q = m$. (Here the metric is induced from the standard embedding of $S^p\left(\sqrt{\frac{p}{m}}\right) \times S^q\left(\sqrt{\frac{q}{m}}\right)$ in $S^{m+1}(1)$.)
Or
(b) the Veronese surface in $S^4(1)$; see VIII (1.20).

The first assertion is due to Simons [Si]; and the second to Lawson [L1] in case $m = n - 1$; and to Chern–do Carmo–Kobayashi [CCK] for arbitrary codimension.

CHAPTER II. IMMERSIONS OF PARALLEL MEAN CURVATURE

INTRODUCTION

We begin with the first properties of Riemannian immersions $\varphi : M \to N$ of parallel mean curvature. In particular, Theorem (1.10) of Ruh–Vilms characterizes such $\varphi : M \to \mathbf{R}^n$ in terms of harmonicity of its Gauss map $\gamma : M \to G_{n,m}$. Hypersurfaces of parallel mean curvature are characterized as extrema of an isoperimetric variational principle in Theorem (1.16).

In Section 2 we reproduce Alexandrov's characterization (Theorem (2.6)) of the compact embedded hypersurfaces of $\mathbf{R}^n$ of constant mean curvature (c.m.c.).

1. PARALLEL MEAN CURVATURE

(1.1) The *mean curvature* of a Riemannian immersion $\varphi : M \to N$ is the vector field $\tau(\varphi)/m$ along φ. From I (1.16) we observe that $\tau(\varphi)/m$ *is everywhere orthogonal to* M.

(1.2) Say that φ has *parallel mean curvature* if $\tau(\varphi)$ is a covariant constant; i.e.,

$$\nabla^\perp \tau(\varphi) \equiv 0$$

as a section of the normal bundle of M in N.
Because $\nabla_X |\tau(\varphi)|^2 = 2 < \nabla_X^\perp \tau(\varphi), \tau(\varphi) >$ we see that $|\tau(\varphi)|$ is constant. If $n - m = 1$, then $\nabla^\perp \tau(\varphi) \equiv 0$ iff $|\tau(\varphi)|$ is constant; in that case we shall say that φ has *constant mean curvature*.

(1.3) **Proposition.** *Let* $\varphi : M^m \to S^{n-1}$ *be a Riemannian immersion of parallel mean curvature and* $i : S^{n-1} \to \mathbf{R}^n$ *the standard inclusion map. Then* $\Phi = i \circ \varphi : M \to \mathbf{R}^n$ *has parallel mean curvature; and conversely.*

Proof. $\nabla^\perp \tau(i \circ \varphi) = \nabla^\perp(di \cdot \tau(\varphi)) + \nabla^\perp \text{ Trace } \nabla di(d\varphi, d\varphi)$

Now

$$\nabla_X^\perp(di \cdot \tau(\varphi)) = (\nabla_{di \cdot d\varphi(X)}^{\mathbf{R}^n}(di \cdot \tau(\varphi)))^\perp = (\nabla_{d\varphi(X)}^{S^{n-1}} \tau(\varphi))^\perp + (\nabla di(d\varphi(X), \tau(\varphi)))^\perp .$$

The first term $= 0$ by hypothesis; the second vanishes because $d\varphi(X) \perp \tau(\varphi)$. Let ν be the outer normal vector field to S^{n-1}. We observe that

$$\text{Trace } \nabla di(d\varphi, d\varphi) = -m\nu$$

because φ is a Riemannian immersion. Now

$$\nabla_X^\perp(-m\nu) = -m\left(\nabla_{d\varphi(X)}\nu\right)^\perp = -m(d\varphi(X))^\perp = 0.$$

Thus $\nabla^\perp \tau(i \circ \varphi) = 0$. The converse statement follows by the same argument. //

(1.4) Proposition. *Let G be a connected compact Lie group of isometries of $\mathbb{R}^n$. Any orbit M of codimension 2 in $\mathbb{R}^n$ has parallel mean curvature. Almost none of those lies minimally in a hypersphere.*

Proof. Orbits are homogeneous spaces, so the tension field τ of M and its centre of mass c are both G-invariant. Consequently, M lies in a hypersphere S centred at c. Both $|\tau|^2$ and $<\tau, v>$ are constant on M, where v denotes the position vector with respect to c.

Now M has codimension 1 in S, so by (1.2) has parallel mean curvature; and by Proposition (1.3) M has parallel mean curvature in $\mathbb{R}^n$. If M lies minimally in some hypersphere $\tilde{S}$, we must have $\tilde{S} = S$; for otherwise $M = \tilde{S} \cap S$ (a hypersphere of $\tilde{S}$), and could be minimal in $\tilde{S}$ only if its centre of mass is the centre of $\tilde{S}$. But M is minimal in S iff it is a V-extremal amongst all orbits of the same type in S (see also V (1.13)). //

(1.5) Let $G_{n,m}$ denote the *Grassmann manifold* of m-spaces through the origin in $\mathbb{R}^n$. It has the homogeneous representation

$$G_{n,m} = 0(n)/0(m) \times 0(n-m).$$

If L is an m-space and $\bar{L}$ its associated point in $G_{n,m}$, the tangent space of $G_{n,m}$ at $\bar{L}$ can be interpreted as the space of linear maps from L to its orthogonal complement $L^\perp$. If $K \to G_{n,m}$ denotes the bundle whose fibre over $\bar{L}$ is L, then the tangent vector bundle of $G_{n,m}$ is $T(G_{n,m}) = K^* \otimes K^\perp$.

(1.6) For a Riemannian immersion $\varphi: M^m \to \mathbb{R}^n$, its *Gauss map*

$$\gamma: M \to G_{n,m}$$

assigns to each point $x \in M$ the tangent space to $\varphi(M)$ at $\varphi(x)$, translated to the origin in $\mathbb{R}^n$.

(1.7) Lemma. *The induced bundle $\gamma^{-1}T(G_{n,m})$ is canonically isometrically isomorphic to $T^*(M) \otimes V(\mathbb{R}^n, M)$, where $V(\mathbb{R}^n, M)$ denotes the normal bundle* of M in $\mathbf{R}^n$. With that identification the bundles have the same metric connection; and

(1.8) $$d\gamma = \nabla \, d\varphi.$$

The required identifications are made explicit in [EL3, Sec.2].

(1.9) From (1.8) we obtain

$$\nabla \, d\gamma = \nabla^\perp \nabla \, d\varphi.$$

With the notation Trace $\nabla_- \nabla_-$ to indicate that the trace is taken on the two marked vectors, we compute

$$\tau(\gamma) = (\text{Trace } \nabla_- \nabla_- d\varphi)^\perp = (\text{Trace } (\nabla \nabla_- d\varphi(-) + R(,-)d\varphi(-)))^\perp,$$

the curvature being that of the bundle $T^*(M) \otimes \varphi^{-1} T(\mathbf{R}^n)$; i.e., minus that of $T(M)$. Therefore

$$\tau(\gamma) = \nabla^\perp \text{Trace } \nabla \, d\varphi + (d\varphi \cdot R^M(-,)-)^\perp = \nabla^\perp \tau(\varphi).$$

because $(d\varphi(X))^\perp = 0$ for all $X \in T(M)$. Thus we obtain the

(1.10) **Theorem** (Ruh–Vilms). *If $\varphi : M \to \mathbf{R}^n$ is a Riemannian immersion, then the tension field of its Gauss map γ is the normal covariant derivative of its tension field:*

$$\tau(\gamma) = \nabla^\perp \tau(\varphi).$$

In particular, φ has parallel mean curvature iff γ is harmonic.

(1.11) In a similar vein, if $\varphi : M^m \to S^n$ is a Riemannian immersion, then Obata's Gauss map

$$\tilde{\gamma} : M^m \to G_{n+1,m+1}$$

assigns to each point $x \in M$ the $(m+1)$-space in $\mathbf{R}^{n+1}$ which intersects S^n in the totally geodesic m-sphere whose tangent space at $\varphi(x)$ is $d\varphi(x)(T_x(M))$. Then

(1.12) **Theorem.** *φ is a minimal immersion iff $\tilde{\gamma}$ is harmonic.*

(1.13) Let $\varphi : M^{n-1} \to N^n$ be an oriented Riemannian immersion of one oriented manifold into another; and η the unit normal vector field along φ such that $T_x(M) \wedge \eta$ determines the orientation of $T_{\varphi(x)}(N)$ for all $x \in M$. Let $\Phi : (-\varepsilon, \varepsilon) \times M \to N$ be a deformation of $\varphi = \Phi(0, \cdot)$ through Riemannian immersions. We write φ_t for $\Phi(t, \cdot)$ and associate to Φ a function $D : (-\varepsilon, \varepsilon) \to \mathbb{R}$ by

(1.14) $$D(t) = \int_{[0,t] \times M} \Phi^* \nu_N$$

where ν_N is the volume form of N.

As in I(2.5) we consider the volume functional $V(t) = V(\varphi_t) = \int_M \nu(t)$ with $\nu(t) = \varphi_t^* \nu_N$, and form the isoperimetric integral in the context of Lagrange multipliers: let

$$W(t) = V(t) + (n-1)H\, D(t),$$

where

(1.15) $$(n-1)H = [V(0)]^{-1} \int_M <\tau(\varphi), \eta> \nu.$$

(1.16) **Theorem.** *For a Riemannian immersion $\varphi : M^{n-1} \to N^n$, the following properties are equivalent*:

(i) *φ has constant mean curvature H*;

(ii) *for all D-preserving variations*, $\left.\dfrac{dV(t)}{dt}\right|_0 = 0$;

(iii) *for all variations*, $\left.\dfrac{dW(t)}{dt}\right|_0 = 0$.

Proof. Step 1.

(1.17) $$\left.\frac{dW(t)}{dt}\right|_0 = \left.\frac{dV(t)}{dt}\right|_0 + (n-1)H \left.\frac{dD(t)}{dt}\right|_0$$

$$= -\int_M <\tau(\varphi), \left.\frac{\partial \varphi_t}{\partial t}\right|_0> \nu + (n-1)H \int_M <\left.\frac{\partial \varphi_t}{\partial t}\right|_0, \eta> \nu.$$

Proof. $\left.\dfrac{dV(t)}{dt}\right|_0$ was computed in I(2.5). We prove

(1.18) $$\left.\frac{dD(t)}{dt}\right|_0 = \int_M <\left.\frac{\partial \varphi_t}{\partial t}\right|_0, \eta> \nu.$$

Let $x \in M$ and choose a positively oriented orthonormal frame $e_1, \ldots, e_{n-1}, e_n = \eta$ around $\varphi(x)$. Then

$$\Phi^* \nu_N = a(t,x) dt \wedge \nu,$$

where $a(t,x) = \Phi^* \nu_N \left(\dfrac{\partial}{\partial t}, e_1, \ldots, e_{n-1}\right) = \nu_N \left(\dfrac{\partial \varphi_t}{\partial t}, d\varphi_t(e_1), \ldots, d\varphi_t(e_{n-1})\right)$

$$= <\dfrac{\partial \varphi_t}{\partial t}, \eta_t>,$$

where η_t is the unit normal associated to the immersion φ_t. Then

$$\left.\dfrac{dD(t)}{dt}\right|_0 = \int_M \left.\dfrac{\partial(\Phi^* \nu_N)}{\partial t}\right|_0 = \int_M a(0,x) \cdot \nu$$

which gives (1.18).

Step 2. Assume that i) holds; then $\tau(\varphi) = (n-1)H\eta$. Thus i) $\Rightarrow$ iii) follows immediately from (1.17).

iii) $\Rightarrow$ ii) is obvious. So it only remains to prove that ii) $\Rightarrow$ i).

Step 3. A variation is *normal* if $\left.\dfrac{\partial \varphi_t}{\partial t}\right|_0$ is everywhere parallel to η. We have

(1.19) Lemma. *For any smooth $f: M \to \mathbb{R}$ such that $\int_M f\nu = 0$, there exists a D-preserving normal variation φ_t with $\left.\dfrac{\partial \varphi_t}{\partial t}\right|_0 = f\eta$.*

Proof. We consider variations of the form

(1.20) $\qquad \Phi(t,x) = \exp_{\varphi(x)}(R(t,x)\eta), \quad t \in (-\varepsilon, \varepsilon), \quad x \in M$

where $R: (-\varepsilon, \varepsilon) \times M$ is a smooth function to be determined. Note that the variation (1.20) is normal, because $\left.\dfrac{\partial \varphi_t}{\partial t}\right|_0 = \left.\dfrac{\partial R}{\partial t}\right|_0 \eta$.

Let us compute the function $D(t)$ associated with (1.20). We observe that $\Phi = e \circ \psi$, where $\psi: (-\varepsilon, \varepsilon) \times M \to \mathbb{R} \times M$ is the map $\psi(t,x) = (R(t,x), x)$; and $e(u,x) = \exp_{\varphi(x)} u\eta$, $u \in \mathbb{R}$.

By setting $E(u,x) = \det(de(u,x))$, we obtain
(1.21)
$$D(t) = \int_{[0,t]\times M} \Phi^* \nu_N = \int_{[0,t]\times M} \psi^* e^* \nu_N$$
$$= \int_{[0,t]\times M} E(R(t,x),x) \dfrac{\partial R}{\partial t}(dt \wedge \nu) = \int_M \left(\int_0^t E(R(t,x),x) \dfrac{\partial R}{\partial t} dt\right) \nu.$$

Now let R be the solution of the initial value problem

$$\dfrac{\partial R}{\partial t} = \dfrac{f(x)}{E(R(t,x),x)} \qquad R(0,x) = 0.$$

We conclude from (1.21) that $D(t) = t \int_M f \, \nu = 0 = D(0)$, so our variation is D-preserving. And $\left.\frac{\partial \varphi_t}{\partial t}\right|_0 = \left.\frac{\partial R}{\partial t}\right|_0 \eta = f \eta$, because $E(R(0,x),x) = 1$. //

Step 4. We prove that ii) $\Rightarrow$ i). Suppose that at a point $x \in M$ we have $\tau(\varphi)(x) \neq (n-1)H \eta$. Writing $\tau(\varphi) = T \eta$, we can assume $T(x) > (n-1)H$.

Set
$$M^+ = \{x' \in M : T(x') > (n-1)H\}, \quad M^- = \{x' \in M : T(x') < (n-1)H\}.$$

Let α^+, α^- be non-negative piecewise smooth functions on M such that

$$x \in \text{supp } \alpha^+ \subset M^+, \text{supp } \alpha^- \subset M^-, \int_M (\alpha^+ + \alpha^-)(T - (n-1)H)\nu = 0$$

where supp $\alpha^\pm$ denotes the support of $\alpha^\pm$. From the definition (1.15) of H, $\int_M (T - (n-1)H)\nu = 0$. Thus α^+ and α^- exist.

Set $f = (\alpha^+ + \alpha^-)(T - (n-1)H)$. By hypothesis (ii)

$$0 = \left.\frac{dV(t)}{dt}\right|_0 = -\int_M \left\langle \tau(\varphi), \left.\frac{\partial \varphi_t}{\partial t}\right|_0 \right\rangle \nu = -\int_M T f \nu$$

for the variation φ_t of Lemma (1.19). Thus

$$0 = \int_M f(T - (n-1)H)\nu = \int_M (\alpha^+ + \alpha^-)(T - (n-1)H)^2 \nu > 0,$$

a contradiction. Thus $T \equiv (n-1)H$. //

2. ALEXANDROV'S THEOREM

(2.1) Let f be a function on an open set $U \subset \mathbb{R}^m$. We consider its graph $\Gamma f = \{(x, f(x)) : x \in U\} \subset \mathbb{R}^{m+1}$. Γf has constant mean curvature H iff

$$(2.2) \qquad \sum_{i,j=1}^m A_{ij} f_{ij} = mH(1 + |\nabla f|^2)^{3/2}$$

where $A_{ij} = (1 + |\nabla f|^2)\delta_{ij} - f_i f_j$ and $f_i = \frac{\partial f}{\partial x_i}$, $f_{ij} = \frac{\partial^2 f}{\partial x_i \partial x_j}$.

We observe that the matrix (A_{ij}) is positive definite at each point of U; so (2.2) is elliptic.

(2.3) **Lemma.** *Let $f, \tilde{f} : U \to \mathbb{R}$ be solutions of (2.2). Then $u = \tilde{f} - f$ satisfies a linear elliptic equation of the form*

$$(2.4) \qquad \sum_{i,j=1}^m a_{ij} u_{ij} + \sum_{k=1}^m b_k u_k = 0.$$

Proof. Write (2.2) in the form $F(f_{ij}, f_k) = 0$. Then $0 = F(\tilde{f}_{ij}, \tilde{f}_k) - F(f_{ij}, f_k) =$

$$\int_0^1 \frac{d}{d\tau} F(\cdot) d\tau = \int_0^1 \left[\sum_{i,j=1}^m (\tilde{f}_{ij} - f_{ij}) \frac{\partial F}{\partial f_{ij}}(\cdot) + \sum_{k=1}^m (\tilde{f}_k - f_k) \frac{\partial F}{\partial f_k}(\cdot) \right] d\tau,$$

where $(\cdot) = (\tau \tilde{f}_{ij} + (1 - \tau) f_{ij}, \tau \tilde{f}_k + (1 - \tau) f_k)$.

Setting

$$a_{ij} = \int_0^1 \frac{\partial F}{\partial f_{ij}}(\cdot) d\tau, \quad b_k = \int_0^1 \frac{\partial F}{\partial f_k}(\cdot) d\tau$$

gives (2.4). Ellipticity follows because $\partial F / \partial f_{ij} = A_{ij}$: for any $0 \neq \xi \in \mathbb{R}^m$,

$$a_{ij} \xi^i \xi^j = \int_0^1 A_{ij}(\cdot) \xi^i \xi^j \, d\tau > 0. \; //$$

We shall use the following characterization of spheres:

(2.5) Lemma. *Let $M \hookrightarrow \mathbb{R}^{m+1}$ be a compact hypersurface having a hyperplane of symmetry in each direction. Then M is a Euclidean sphere.*

Proof. Take $m + 1$ mutually orthogonal hyperplanes of symmetry $P_1, \ldots, P_{m+1}$; and let C be their unique common point. We shall show that M is invariant under all rotations about C.

Let P be any other hyperplane of symmetry of M. Then $C \in P$; for otherwise suitable compositions of reflections through $P_1, \ldots, P_{m+1}, P$ will take any given point of M to a point arbitrarily far from C, contradicting the compactness of M.

Thus M is invariant under reflection through every hyperplane containing C. The assertion follows, because every rotation about C is a composition of such reflections. //

(2.6) Theorem (Alexandrov). *Let M^m be a compact manifold embedded in $\mathbb{R}^{m+1}$ with constant mean curvature. Then M^m is a Euclidean sphere ($m \geq 2$).*

Proof. First of all, M is the boundary of a compact domain $D \subset \mathbb{R}^{m+1}$. We choose cartesian coordinates on $\mathbb{R}^{m+1}$ with given x^{m+1}-direction, so that D lies in the region $x^{m+1} \geq 0$ and touches the hyperplane $x^{m+1} = 0$. For each $a > 0$ let P_a be the hyperplane $x^{m+1} = a$. Let M_a be the part of M which lies in the region

$x^{m+1} \leq a$; and $\widetilde{M}_a$ its reflection across P_a.

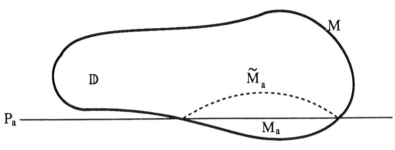

Figure 2.7

Clearly $\widetilde{M}_a$ lies inside D for sufficiently small a. Define

$$c = \sup\{b \in \mathbb{R}(>0) : \widetilde{M}_a \subset D \quad \text{for all} \quad 0 < a \leq b\}.$$

Then $c < +\infty$, because M is compact. We shall show that P_c is a hyperplane of symmetry of M.

There are two eventualities:

Case 1. There is a point $x_0 \in M \cap \widetilde{M}_c$ not belonging to P_c.

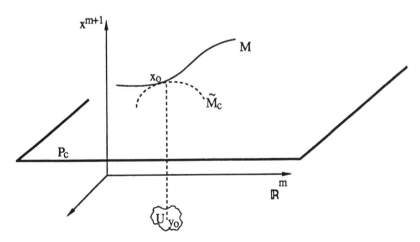

Figure 2.8

Now there is a sufficiently small open neighbourhood $U \subset \mathbb{R}^m$ of $y_0 = (x_0^1, \ldots, x_0^m)$ on which the hypersurfaces M and $\widetilde{M}_c$ (both having constant mean curvature H)

can be represented as the graphs of functions f and $\tilde{f}$, respectively, as in Lemma (2.3); and $f(y_0) = \tilde{f}(y_0) = x_0^{m+1}$. Clearly $u = \tilde{f} - f \leq u(y_0) = 0$. Because u satisfies (2.4) we can apply the maximum principle I(1.18) to conclude that $f = \tilde{f}$ in a neighbourhood of y_0. I.e., M and $\widetilde{M}_c$ coincide near x_0.

Case 2. $M \cap \widetilde{M}_c \subset P_c$.

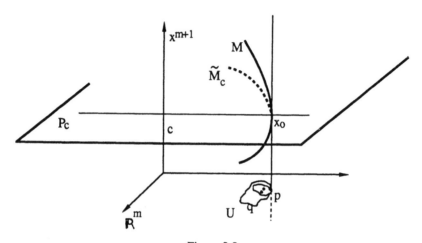

Figure 2.9

This case is handled similarly: We have a point $x_0 \in P_c$ such that the line through x_0 parallel to the x^{m+1}-axis belongs to $T_{x_0}(M)$. That guarantees that the hypotheses of I(1.18), Case 2 are satisfied (with U, p and q as in Fig.2.9). Thus we conclude that $M \cap \{x^{m+1} \geq c\}$ and $\widetilde{M}_c$ coincide near x_0.

In either case, to show that P_c is a hyperplane of symmetry we proceed as follows: If $\tilde{N}_c$ is the component of $\widetilde{M}_c$ containing x_0, then $\tilde{N}_c \subset M$. Indeed, $\tilde{N}_c \cap M$ is closed in M, and the above arguments show that it is also open. It is clear that $\tilde{N}_c$ is the reflection across P_c of a piece N_c of M_c; and that $N_c \cup \tilde{N}_c$ is a closed m–dimensional topological submanifold of M. Therefore $N_c \cup \tilde{N}_c = M$.

Finally, because the x^{m+1}-direction was chosen arbitrarily, we conclude that M has a hyperplane of symmetry in each direction; thus the theorem follows from Lemma (2.5).

3. Notes and comments

(3.1) Matsushima [M] has interpreted the second fundamental form $\beta(\varphi) = \nabla d\varphi$ and the mean curvature $\tau(\varphi)/m$ of a Riemannian immersion in terms of differential forms. If $N = N(c)$ is a space form, then $d\beta(\varphi) = 0$ is Codazzi's equation; and $d^*\beta(\varphi) = 0$ iff M has parallel mean curvature.

(3.2)   Theorem (1.10) is due to Ruh and Vilms [RV]. Theorem (1.12) was found by T. Ishihara [I2] and X-P. Chen [XPC].

(3.3)   Proposition (1.4) is due to Smyth [Smy1]. He gives an example to show that it fails in higher codimension.

(3.4)   The isoperimetric variational principle in Theorem (1.16) characterizing hypersurfaces of constant mean curvature can be found in [BdC] and [BdCE]. It has a higher codimensional analogue [G]; but that is not as useful.

(3.5)   An alternative proof of Alexandrov's theorem [Al] has been given by Reilly [Re] – based on integral formulas.

(3.6)   Alexandrov's proof works in $\mathbb{R}^{m+1}$, H^{m+1}, or an open hemisphere of S^{m+1}. See [Al] and [dCL]. By way of contrast, there are examples of complete constant mean curvature space–like hypersurfaces in Minkowski space L^{n+1} for $n \geq 2$, which are not congruent to the hyperbolic space $H^n = \{x \in L^{n+1} :< x, x > = -1\}$ [Tr].

CHAPTER III. SURFACES OF PARALLEL MEAN CURVATURE

INTRODUCTION

In the variational theory of Riemannian immersions, there are many special features in case the domain M is a Riemann surface:

(1) The energy functional is invariant under conformal changes of the metric on M. Moreover, we have a holomorphic quadratic differential associated with any harmonic map $\varphi : M \to (N, h)$ (Proposition (1.7)). That is obtained by taking the induced tensor field $\varphi^* h$ on M, and extracting the dz^2-part of its complexification. Thus φ is a (weakly) conformal map iff $(\varphi^* h)^{2,0} \equiv 0$. Through that differential we can characterize conformal harmonic maps $\varphi : M \to (N, h)$ as minimal branched immersions.

(2) The Grassmannian $G^0_{n,2}$ of oriented 2–spaces in $\mathbf{R}^n$ is identified with the complex quadric. In addition to Ruh–Vilms' Theorem (II (1.10)) – valid now for conformal immersions – we have Chern's Theorem (1.15), that *such a $\varphi : M \to \mathbf{R}^n$ is harmonic iff its Gauss map is anti–holomorphic.*

(3) Let $\varphi : M \to N^n(c)$ be a conformal immersion into a space form, and β_φ its second fundamental form. *If φ has parallel mean curvature, then $\beta^{2,0}_\varphi$ is holomorphic* (Proposition 2.8 of H. Hopf).

That latter is used in the following three results: *Let M be a compact Riemann surface of genus* 0.

Theorem of Almgren–Calabi (2.9): *Any harmonic map $\varphi : M \to S^3$ has image contained in a totally geodesic hypersphere.*

Theorem of Hopf (2.10): *If $\varphi : M \to \mathbf{R}^3$ is an isometric immersion of constant mean curvature, then $\varphi(M)$ is an embedded Euclidean sphere.*

The immersions $\varphi : \mathbf{R}^2 \to \mathbf{R}^3$ of constant mean curvature are characterized as solutions of the sinh–Gordon equation (3.3). Certain doubly periodic solutions determine *immersions of constant mean curvature of a 2–torus in $\mathbf{R}^3$* (Theorem (4.8) of Wente, in a special form found by Abresch).

1. THEOREMS OF CHERN AND RUH–VILMS

(1.1) Let M be a Riemann surface; i.e., an oriented 2–dimensional manifold equipped with a conformal equivalence class of Riemannian metrics. Let (N, h)

be a Riemannian manifold with metric h, sometimes denoted by $<,>$; its complex bilinear extension to the complexified tangent bundle $T^\mathbb{C}(N) = T(N) \otimes_\mathbb{R} \mathbb{C}$ is denoted by $<,>^\mathbb{C}$.

For a map $\varphi : M \to (N, h)$ we have

(1.2)
$$\partial_z \varphi = \varphi_z = \frac{1}{2}(\varphi_x - i\varphi_y)$$
$$\partial_{\bar{z}} \varphi = \varphi_{\bar{z}} = \frac{1}{2}(\varphi_x + i\varphi_y)$$

in an isothermal chart $z = x + iy$ of M. In particular, φ is harmonic iff

(1.3) $\quad \nabla_{\bar{z}} \partial_z \varphi = 0;\quad$ or equivalently by (2.3) below, $\quad \nabla_z \partial_{\bar{z}} \varphi = 0$,

where ∇ is the covariant derivative in the bundle $\varphi^{-1} T^\mathbb{C}(N)$.

In charts, (1.3) takes the form

(1.4) $\quad \varphi^\gamma_{z\bar{z}} + {}^N\Gamma^\gamma_{\alpha\beta} \varphi^\alpha_z \varphi^\beta_{\bar{z}} = 0 \quad (1 \leq \gamma \leq n = \dim N)$.

(1.5) A map $\varphi : M \to (N, h)$ is *(weakly) conformal* if its first fundamental form $\varphi^* h$ has complexification with vanishing $(2,0)$-part:

(1.6) $\quad (\varphi^* h)^{2,0} = <\varphi_z, \varphi_z>^\mathbb{C} dz^2 \equiv 0$.

Indeed,
$$<\varphi_z, \varphi_z>^\mathbb{C} = \frac{1}{4}\left(|\varphi_x|^2 - |\varphi_y|^2 - 2i <\varphi_x, \varphi_y>\right).$$

(1.7) **Proposition.** *If $\varphi : M \to (N, h)$ is a harmonic map, then $(\varphi^* h)^{2,0}$ is a holomorphic quadratic differential.*

Proof.
$$\partial_{\bar{z}} <\varphi_z, \varphi_z>^\mathbb{C} = 2 <\nabla_{\bar{z}} \varphi_z, \varphi_z>^\mathbb{C} \equiv 0$$

by (1.3). //

(1.8) **Application.** *Any harmonic map $\varphi : S^2 \to (N, h)$ is weakly conformal.* Indeed, the Riemann–Roch theorem insures that every holomorphic quadratic differential on S^2 vanishes. Furthermore, we note that the holomorphic quadratic differential $(\varphi^* h)^{2,0}$ can be used to prove the following characterization:

(1.9) **Proposition.** *A non-constant map $\varphi : M \to (N, h)$ is a minimal branched immersion iff it is conformal harmonic.*

(1.10) Let $\varphi: M \to \mathbb{R}^n$ be a conformal immersion; and $G^0_{n,2}$ denote the Grassmannian of oriented two planes in $\mathbb{R}^n$ (equivalently, $G^0_{n,2} = SO(n)/SO(2) \times SO(n-2)$).

The Gauss map $\gamma_\varphi : M \to G^0_{n,2}$ assigns to each $x \in M$ the image $d\varphi(x)(T_x(M))$ in $T_{\varphi(x)}(\mathbb{R}^n)$, translated to the origin (see II (1.6)).

(1.11) Each point $\alpha \in G^0_{n,2}$ determines a complex line in $\mathbb{C}^n = \mathbb{R}^n \otimes_\mathbb{R} \mathbb{C}$. That gives a smooth embedding

(1.12) $$j : G^0_{n,2} \to \mathbb{C}P^{n-1}.$$

If α is represented by the exterior product $v \wedge w$ of two orthonormal vectors, then

(1.13) $$j(\alpha) = [z],$$

the complex line determined by $z = v + iw \in \mathbb{C}^n$; and the isotropy condition

$$<z,z>^C = 2i <v,w> = 0$$

displays the image of j as the complex quadric Q_{n-2} in $\mathbb{C}P^{n-1}$, whose equation in homogeneous coordinates $z_1, \ldots, z_n$ is $\sum_{k=1}^n z_k^2 = 0$. By way of summary:

(1.14) **Lemma.** *There is a canonical identification* $j: G^0_{n,2} \xrightarrow{\simeq} Q_{n-2}$, *the quadric hypersurface of* $\mathbb{C}P^{n-1}$.

Using that identification, we have

$$\partial_{\bar{z}}[\gamma_\varphi] = [\partial_z \partial_{\bar{z}} \varphi].$$

Now (1.3) gives

(1.15) **Theorem** (Chern). *The map j in (1.13) identifies γ_φ with $[\partial_{\bar{z}} \varphi]$. A conformal immersion $\varphi : M \to \mathbb{R}^n$ is harmonic iff $\gamma_\varphi : M \to Q_{n-2}$ is anti–holomorphic.*

(1.16) Let β_φ denote the second fundamental form (with values in the normal bundle). As in (1.5), its complexification has type decomposition

$$\beta^C_\varphi = \beta^{2,0}_\varphi + \beta^{1,1}_\varphi + \beta^{0,2}_\varphi.$$

The *umbilic points* of φ are the zeros of $\beta_\varphi^{2,0}$. Say that a conformal immersion $\varphi : M \to \mathbf{R}^n$ is *totally umbilic* if $\beta_\varphi^{2,0} \equiv 0$.

Now identify $d\gamma_\varphi = \beta_\varphi$, as in II (1.8). A simple calculation (using $\nabla_z \partial_{\bar z} \varphi - \nabla_{\bar z} \partial_z \varphi = 0$, in the notation of (2.2) below) shows that

$$\beta_\varphi^C = \partial_z \gamma_\varphi + \partial_{\bar z} \gamma_\varphi .$$

Thus β_φ^C has type $(1,1)$ – i.e., $\beta_\varphi^{2,0} \equiv 0 \equiv \beta_\varphi^{0,2}$ iff $\partial_{\bar z} \gamma_\varphi = 0$. Otherwise said:

(1.17) **Proposition.** *A conformal immersion $\varphi : M \to \mathbf{R}^n$ has holomorphic Gauss map iff it is totally umbilic.*

Examination of Theorem II (1.10) shows that for 2–dimensional domains

(1.18) **Theorem.** *A conformal immersion $\varphi : M \to \mathbf{R}^n$ has parallel mean curvature iff the Gauss map $\gamma_\varphi : M \to Q_{n-2}$ is harmonic.*

(1.19) Let $\varphi : M \to S^n$ be a conformal immersion. And let $\Phi = i \circ \varphi$, where $i : S^n \to \mathbf{R}^{n+1}$ is the standard inclusion. Then Obata's Gauss map

(1.20) $$\gamma_\varphi = \Phi \wedge \gamma_\Phi ;$$

and φ is harmonic iff γ_φ is, by Theorem II (1.12).

2. THEOREMS OF ALMGREN–CALABI AND HOPF

(2.1) We write the complexified differential $d^C = \partial_z + \partial_{\bar z}$. Similarly for ∇.

(2.2) **Lemma.** *For any map $\varphi : M \to (N, h)$ we have*

$$\nabla_{\bar z} \partial_z \varphi = \nabla_z \partial_{\bar z} \varphi .$$

Proof. $\nabla_{\bar z} \partial_z \varphi - \nabla_z \partial_{\bar z} \varphi = [\partial_z \varphi, \partial_{\bar z} \varphi] = \varphi_*[\partial_z, \partial_{\bar z}] = 0$. //

Similarly,

(2.3) **Lemma.** *φ is harmonic iff $\nabla_z \partial_{\bar z} \varphi = 0$. Equivalently $\nabla_{\bar z} \partial_z \varphi = 0$.*

(2.4) In an isothermal chart with $z = x+iy$, we can write the second fundamental form
$$\beta_\varphi = B_{11}\, dx^2 + 2B_{12}\, dx\, dy + B_{22}\, dy^2.$$

Its complexification
$$\beta^C_\varphi = \nabla^C d^C\varphi = \beta^{2,0}_\varphi + \beta^{1,1}_\varphi + \beta^{0,2}_\varphi,$$

so locally

(2.5) $\quad \begin{cases} \beta^{2,0}_\varphi &= \tfrac{1}{4}(B_{11} - B_{22} - 2iB_{12})dz^2 \\ \beta^{1,1}_\varphi &= \tfrac{1}{2}(B_{11} + B_{22})\, dz\, d\bar{z} \\ \beta^{0,2}_\varphi &= \tfrac{1}{4}(B_{11} - B_{22} + 2iB_{12})d\bar{z}^2 \end{cases}$

(2.6) We specialize to the case of a space form $N = N^n(c)$.

Codazzi's equation is
$$(\nabla_X \beta_\varphi)(Y, Z) - (\nabla_Y \beta_\varphi)(X, Z) = 0.$$

Now writing $\beta = \beta_\varphi$, $\nabla_X(\beta(Y, Z)) = (\nabla_X \beta)(Y, Z) + \beta(\nabla_X Y, Z) + \beta(Y, \nabla_X Z)$.
Therefore
$$\nabla_X(\beta(Y, Z)) - \nabla_Y(\beta(X, Z)) = \beta([X, Y], Z) + \beta(Y, \nabla_X Z) - \beta(X, \nabla_Y Z).$$

Take $X = \partial_x, Y = \partial_y$; then take $Z = \partial_x$ and ∂_y:
$$\begin{cases} \nabla_X B_{12} - \nabla_Y B_{11} = \beta(\partial_y, \nabla_{\partial_x}\partial_x) - \beta(\partial_x, \nabla_{\partial_y}\partial_x) = 0 \\ \nabla_X B_{22} - \nabla_Y B_{12} = \beta(\partial_y, \nabla_{\partial_x}\partial_y) - \beta(\partial_x, \nabla_{\partial_y}\partial_y) = 0. \end{cases}$$

In that form $\nabla \tau(\varphi) = 0$ iff
$$\begin{cases} \nabla_{\partial_x} B_{11} + \nabla_{\partial_x} B_{22} = 0 \\ \nabla_{\partial_y} B_{11} + \nabla_{\partial_y} B_{22} = 0, \quad \text{so together we obtain} \end{cases}$$

(2.7) $\quad \begin{cases} \nabla_X B_{12} - \nabla_Y \left(\dfrac{B_{11} - B_{22}}{2}\right) = 0 \\ \nabla_Y B_{12} + \nabla_X \left(\dfrac{B_{11} - B_{22}}{2}\right) = 0. \end{cases}$

I.e., $\nabla_{\bar{z}}\beta^{2,0}_\varphi \equiv 0$. Thus either φ is totally umbilic, or its umbilic points are isolated. Also,

(2.8) Proposition. *If $\varphi : M \to N^n(c)$ is a conformal immersion with parallel mean curvature, then $\beta_\varphi^{2,0}$ is holomorphic.*

That is especially interesting for $n = 3$; for then we can identify $\beta_\varphi^{2,0}$ with a holomorphic quadratic differential on M. These all vanish on S^2, so

(2.9) Corollary (Almgren–Calabi). *If M is a compact oriented surface of genus 0, then any harmonic map $\varphi : M \to S^3$ factors through a totally geodesic map of an equatorial 2-sphere into S^3.*

Proof. M has a unique conformal structure, and by (1.8) φ is weakly conformal. However, $\beta_\varphi^{2,0} \equiv 0 \equiv \beta_\varphi^{0,2}$; and also $\beta_\varphi^{1,1} \equiv 0$ because φ is harmonic. We conclude that $\beta_\varphi \equiv 0$. Now the assertion follows readily. //

(2.10) Theorem (H. Hopf). *Let $\varphi : M \to \mathbb{R}^3$ be an isometric immersion of a closed oriented surface of genus 0. If φ has constant mean curvature, then $\varphi(M)$ is an embedded Euclidean sphere.*

Proof. First of all, its Gauss map $\gamma_\varphi : M \to S^2$ is holomorphic ($\partial_{\bar{z}} \gamma_\varphi \equiv 0$): For Theorem (1.18) says that γ_φ is harmonic; and any harmonic map between such surfaces is $\pm$ holomorphic. However, if γ_φ were – holomorphic, then φ would be harmonic, by Theorem (1.15). But that is impossible by the maximum principle.

Next, γ_φ is holomorphic iff φ is totally umbilic; indeed, $\beta_\varphi^c = \partial_z \gamma_\varphi + \partial_{\bar{z}} \gamma_\varphi$, which has type (1, 1) iff $\partial_{\bar{z}} \gamma_\varphi = 0$. Now φ maps M isometrically and bijectively onto some Euclidean 2-sphere using [KN II, Ch.VII, Theorem 5.1].

(2.11) Proposition. *If M is a closed oriented surface of genus 0 and $\varphi : M \to \mathbb{R}^n$ an isometric immersion with parallel mean curvature ($|\tau(\varphi)| = 1$ say), then $\varphi(M)$ is a minimal surface in some Euclidean sphere.*

Proof. $< \beta_\varphi^{2,0}, \tau(\varphi) > \equiv 0$ because $\beta_\varphi^{2,0}$ is holomorphic and $\tau(\varphi)$ parallel. Therefore the principal curvatures in the $\tau(\varphi)$ direction are both $= |\tau(\varphi)| = 1$. Consider the map $F : M \to \mathbb{R}^n$ defined by $F(x) = \varphi(x) - \tau(\varphi)(x)$. Then $dF \equiv 0$, so $F = c$ (constant). I.e., $\varphi(x) = c + \tau(\varphi)(x)$, so $\varphi(M) \subset S^{n-1}(c, 1)$. Furthermore, $\tau(\varphi) \perp S^{n-1}(c, 1)$, so $\varphi(M)$ is minimal in $S^{n-1}(c, 1)$, by I (1.16). //

(2.12) Proposition. *Let $\varphi : M \to \mathbb{R}^n$ be an immersion of a compact manifold. Then, up to translation, $\varphi(M)$ lies in the subspace spanned by $\tau(\varphi)$; and in no smaller one.*

Proof. We can assume $0 \in \varphi(M)$. Let V = least subspace containing $\varphi(M)$; and W the space spanned by $\tau(\varphi)$. Then $W \subset V$. Suppose $W \neq V$. Then there is a non-zero $v \in V$ such that $<\tau(\varphi), v> \equiv 0$. But $-\Delta <\varphi, v> = <\tau(\varphi), v> \equiv 0$, so $<\varphi, v>$ is constant – and hence 0, because $0 \in \varphi(M)$; that contradicts the choice of v. //

We use that to generalize Hopf's Theorem (2.10):

(2.13) Theorem. *Let M be a compact oriented surface of genus 0 immersed in a space form $N(c)$ with parallel mean curvature $\neq 0$. Then M lies minimally immersed in a hypersphere S of constant positive curvature. If dim $N(c) = 4$, then M is embedded as a great hypersphere of S.*

Proof. We take $N(c) = S^n$, for simplicity; the other cases are handled similarly. Composing

$$M \xrightarrow{\varphi} S^n$$
$$\Phi \searrow \quad i \downarrow$$
$$\mathbf{R}^{n+1}$$

and using Propositions (2.11) and (2.12) we obtain the first assertion. Secondly, in case $n = 4$ we have

$$M \xrightarrow{\varphi} S^3$$
$$\searrow \quad \downarrow$$
$$S^4.$$

That φ is an embedding onto an equator follows now from Corollary (2.9). //

We have also

(2.14) Proposition. *Let P denote the 2-dimensional real projective plane. Then every harmonic map $\psi : P \to S^3$ is constant.*

Proof. Denote by $\pi : S \to P$ the two-leaved covering map; and let $\varphi = \psi \circ \pi$. Then φ is harmonic, so by (2.9) ψ maps P to an equatorial 2-sphere. Because φ is weakly conformal by (1.8), it is $\pm$ holomorphic onto its image. But clearly φ has degree zero and therefore is constant.

3. ON THE SINH–GORDON EQUATION

(3.1) Fix a positive number H which we shall normalize (by orientation and scaling) to be $1/2$. And let $\mathbf{M}$ denote the set of complete Riemannian metrics g on $\mathbf{R}^2$

(1) whose Gauss curvature functions K satisfy $-\infty < K < 1/4$ and are bounded away from those extremes; and

(2) which admit an isometric immersion $\varphi : (\mathbb{R}^2, g) \to \mathbb{R}^3$ with constant mean curvature $H = 1/2$. (Such an immersion is determined up to motions in $\mathbb{R}^3$ and its S^1-family of associates, which are parametrized by an angle γ; see Step 3 below.)

(3.2) Proposition. *The correspondence $g \mapsto \omega = \log|\beta_\varphi^{2,0}|$ is a bijection between the metrics in $\mathbf{M}$ and the bounded solutions in $\mathbb{R}^2$ of the sinh–Gordon equation*

(3.3) $$\Delta \omega = \sinh \omega .$$

Here $-\Delta = \partial_x^2 + \partial_y^2 = 4\partial_{z\bar{z}}^2$.

Proof. We work in an isothermal chart, so $g = \rho^2 \, dz \, d\bar{z}$, and Gauss' equation

(3.4) $$K = \frac{B_{11}B_{22} - B_{12}^2}{\rho^4} = \frac{1}{2\rho^2} \Delta \log \rho^2 .$$

Step 1. Away from the umbilics, $g_0 = 4|\beta_\varphi^{2,0}| dz \, d\bar{z}$ is a flat metric. Indeed, by (3.4) its curvature

$$K_0 = \frac{1}{8|\beta_\varphi^{2,0}|} \cdot \Delta \log 4|\beta_\varphi^{2,0}| \equiv 0$$

because $\beta_\varphi^{2,0}$ is holomorphic.

Step 2. From (2.4) and (2.5) we find

$$16|\beta_\varphi^{2,0}|^2 = 4\rho^4 \left[\left(\frac{B_{11} + B_{22}}{2\rho^2}\right)^2 - \left(\frac{B_{11}B_{22} - B_{12}^2}{\rho^4}\right)\right]$$
$$= 4\rho^4(H^2 - K) = \rho^4(1 - 4K) .$$

Because $H^2 - K$ is bounded away from zero, g_0 is a complete flat metric on all $\mathbb{R}^2$; thus we identify $(\mathbb{R}^2, g_0)$ with $\mathbb{C}$, and so have global isothermal coordinates with respect to which $g_0 = \rho^2\sqrt{1 - 4K} \, dz \, d\bar{z}$; therefore

(3.5) $$\rho^2\sqrt{1 - 4K} = 1 .$$

Setting $\omega = \log \rho^2$ and using (3.4) and (3.5) show that ω is bounded, and

$$e^{-\omega}\Delta\omega/2 = K = (1 - e^{-2\omega})/4 ,$$

which establishes (3.3).

Step 3. Conversely, if $\omega : \mathbf{R}^2 \to \mathbf{R}$ is a bounded solution of (3.3), then $g = e^\omega \, dz \, d\bar{z}$ is a complete metric on $\mathbf{R}^2$. And the matrix (B_{ij}) with entries

(3.6)
$$B_{11} = e^{\omega/2}[(\sinh \omega/2) \cos^2 \gamma + (\cosh \omega/2) \sin^2 \gamma]$$
$$B_{12} = -\sin \gamma \cos \gamma = B_{21}$$
$$B_{22} = e^{\omega/2}[(\cosh \omega/2) \cos^2 \gamma + (\sinh \omega/2) \sin^2 \gamma],$$

determines a second fundamental form β satisfying Codazzi's and Gauss' equations (2.6) and (3.4), by (3.3).

The fundamental theorem of surface theory insures that there is an isometric immersion $\varphi : \mathbf{R}^2 \to \mathbf{R}^3$ with induced metric g and second fundamental form β. That has constant mean curvature $\frac{1}{2}$, and its curvature is bounded away from $-\infty$ and $1/4$. //

4. WENTE'S THEOREM

(4.1) Given a conformal immersion $\varphi : \mathbf{R}^2 \to \mathbf{R}^3$ of constant mean curvature $1/2$; assume that φ is doubly periodic – with fundamental domain a rectangle with edges parallel to the axes of $\mathbf{R}^2$. Then $\beta_\varphi^{2,0}$ is itself doubly periodic, and so constant, by Proposition (2.8).

Note that this situation corresponds to the case $\gamma = 0$ in (3.6).

The principal curvatures κ_1, κ_2 of the immersion φ are the eigenvalues of β_φ with respect to the metric induced by φ. These are distinct (i.e., the surface has no umbilics, by Codazzi's equation), and we will label them so that $\kappa_1 < \kappa_2$. Their associated eigenvectors determine two families of curves – which in this case are orthogonal straight lines parallel to the edges of the fundamental domain – called the *lines of curvature*. They are given by

(4.2)
$$(B_{11} - B_{22} - 2iB_{12})dz^2 - (B_{11} - B_{22} + 2iB_{12})d\bar{z}^2 = 0.$$

(4.3) Define the function $\omega : \mathbf{R}^2 \to \mathbf{R}$ by $e^\omega(\kappa_2 - \kappa_1) = 1$. Together with $\kappa_2 + \kappa_1 = 1$ we obtain

(4.4)
$$\kappa_1 = e^{-\omega/2} \sinh \omega/2, \quad \kappa_2 = e^{-\omega/2} \cosh \omega/2.$$

The induced metric is $g = e^{\omega/2} dz \, d\bar{z}$; and of course (3.3) is satisfied, because it is equivalent to Gauss' equation.

(4.5) **Lemma.** ω *is doubly periodic.*

Proof. Fix y and write $\omega(x) = \omega(x,y)$. We can assume that ω is non-trivial. Then there is no point x at which $\omega(x) = 0 = \omega'(x)$.

Take a critical point x_0 of ω. Then either $\omega(x_0) > 0$ or $\omega(x_0) < 0$. The sinh–Gordon equation (3.3) insures that $\omega''(x_0) < 0$ or $\omega''(x_0) > 0$, so that all critical points of ω are isolated and alternate with positive maxima following negative minima.

Let x_0 and $x_0 + p$ be successive maxima of ω. Now the function

$$x \to \frac{\omega'(x)^2}{2} + \cosh \omega(x)$$

is constant, so that $\omega(x_0 + p) = \omega(x_0)$ and $\omega'(x_0 + p) = \omega'(x_0)$. But $\omega_p(x) = \omega(x+p)$ is a solution of (3.3) with the same initial conditions as ω at x_0, so $\omega(x+p) \equiv \omega(x)$.

Similarly for periodicity in y. //

(4.6) We proceed to the inverse construction. To simplify notation, we replace ω by 2ω in (3.3) to obtain

(3.3') $$\Delta \omega = \sinh \omega \cosh \omega .$$

Now, starting with a solution $\omega : \mathbf{R}^2 \to \mathbf{R}$ of (3.3) which is doubly periodic as above, we construct a map $\varphi : \mathbf{R}^2 \to \mathbf{R}^3$ which is conformal:

$$<\varphi_x, \varphi_y> = 0, \quad |\varphi_x|^2 = e^{2\omega} = |\varphi_y|^2$$

and such that

$x \to \varphi(x,y)$ parametrizes the κ_1–curvature lines;

$y \to \varphi(x,y)$ parametrizes the κ_2–curvature lines, where $\kappa_1 = e^{-\omega} \sinh \omega$, $\kappa_2 = e^{-\omega} \cosh \omega$, so φ has constant mean curvature $1/2$.

It is characterized by the system

(4.7)
$$\begin{cases} \varphi_{xx} = \omega_x \varphi_x - \omega_y \varphi_y + e^\omega \sinh \omega \cdot \nu \\ \varphi_{xy} = \omega_y \varphi_x + \omega_x \varphi_y \\ \varphi_{yy} = -\omega_x \varphi_x + \omega_y \varphi_y + e^\omega \cosh \omega \cdot \nu \\ \nu_x = -e^{-\omega} \sinh \omega \, \varphi_x \\ \nu_y = -e^{-\omega} \cosh \omega \, \varphi_y , \end{cases}$$

ν being the unit normal field.

(4.8) **Theorem** (Wente). *There are conformal immersions $\varphi : \mathbf{R}^2 \to \mathbf{R}^3$ with constant mean curvature $H \neq 0$, which are doubly periodic with respect to a rectangle.*

Thus we obtain tori immersed in $\mathbf{R}^3$ with constant mean curvature with respect to the induced metrics.

We shall deal with a very special case, identified by Abresch [Ab2]: One whose κ_1–curvature lines are planar. Their curvature and torsion are related by

$$\kappa_1^2 \cdot \tau_1 = e^{-3\omega}(\sinh\omega \cdot \omega_{xy} - \cosh\omega \cdot \omega_x \omega_y).$$

Therefore we seek solutions of the pair

(4.9) $\quad \begin{cases} \Delta\omega = \sinh\omega \cosh\omega \\ \sinh\omega \cdot \omega_{xy} - \cosh\omega \cdot \omega_x \omega_y = 0 \end{cases}$

Setting $W = \cosh\omega$ transforms (4.9) into

(4.10) $\quad \begin{cases} (W^2 - 1)\Delta W + W|\nabla W|^2 - W(W^2 - 1)^2 = 0 \\ (W^2 - 1)W_{xy} - 2W\, W_x W_y = 0. \end{cases}$

Analysis of (4.10) leads to the following

(4.11) **Proposition.** *Take any real numbers $\alpha, \beta \geq 0$ with $\alpha + \beta \geq 1$. Define the elliptic real valued functions f, g of one variable through*

(4.12) $\quad \begin{aligned} f'^2 &= f^4 + (1 + \alpha^2 - \beta^2)f^2 + \alpha^2, & f(0) &= 0, & f'(0) &= \alpha \\ g'^2 &= g^4 + (1 - \alpha^2 + \beta^2)g^2 + \beta^2, & g(0) &= 0, & g'(0) &= \beta. \end{aligned}$

Then

(4.13) $\quad \begin{cases} \cosh\omega = (f_x + g_y)/(1 + f^2 + g^2), & \omega(0,0) \geq 0 \\ \omega_x = -f(x)\sinh\omega \\ \omega_y = -g(y)\sinh\omega. \end{cases}$

define real analytic solutions ω of (4.9).

Proof. With $W = \cosh\omega$, the system (4.13) becomes

(4.14) $\quad W = \dfrac{f_x + g_y}{1 + f^2 + g^2} \quad$ and

(4.15) $\quad W_x = -f(x)(W^2 - 1), \quad W_y = -g(y)(W^2 - 1).$

But (4.14) implies the first equation in (4.10). And the second is the integrated form

$$\left(\frac{W_y}{W^2 - 1}\right)_x = 0 = \left(\frac{W_x}{W^2 - 1}\right)_y$$

of (4.15).

Direct computation shows that the meromorphic functions f and g described by (4.12) satisfy (4.14) and (4.15). //

(4.16)   Now restrict $\alpha, \beta > 0$ with $\alpha + \beta > 1$ and $\beta - \alpha > 1$. With these assumptions, the meromorphic functions f and g are qualitatively like sin and tan, respectively. Let $a/2$ (resp. b) be the smallest positive zero of f (resp. g). Then the solution ω determined by (4.13) can be extended by reflection; more precisely,

(4.17) **Lemma.**   *Let Γ_{ab} be the group of reflections in $\mathbb{R}^2$ about the lines $x = 0$, $x = a, y = 0, y = b$. Then the solution ω is Γ_{ab}–invariant.*

Theorem (4.8) now follows. //

5.   Notes and comments

(5.1)   Propositions (1.7) and (1.9) are fairly classical (in various degrees of generality). For the latter see [GOR], where a thorough study of branched immersions has been made.

(5.2)   Theorem (1.15) and related aspects of quadric hypersurfaces are found in [Che]. See also [HoO], where Proposition (1.17) appears, as well.

(5.3)   Recognition of the holomorphicity of $\beta^{2,0}(\varphi)$ and its use in proving Theorem (2.10) are due to Hopf [Ho1]. Corollary (2.9) appeared in [Alm] and [Ca]. Proposition (2.11) is [R]'s extension of Hopf's theorem. Theorem (2.13) is due to Hoffman [Hof] and Smyth [Smy1]. For Proposition (2.14) see Lawson [L1].

(5.4)   The relationship between c.m.c. surfaces in $\mathbb{R}^3$ and solutions to the sinh-Gordon equation is classical. We learned of the specific correspondence in Proposition (3.2) in a manuscript of B. Smyth [Smy2].

Pinkall–Sterling [PS] have classified all tori with constant mean curvature in $\mathbb{R}^3$, S^3, H^3– using a reduction of the sinh–Gordon equation (3.3) to a system of O.D.E. Bobenko [Bo] modified that technique to obtain formulas for the immersion in terms of theta functions. See also the detailed constructions by Ercolani, Knörrer, Trubowitz [EKT].

Hitchin [Hi] has given an algebraic geometric approach to the classification of all harmonic maps of a 2–torus into S^3.

(5.5) Wente's theorem (4.8) – in a more general form – appeared in [We1]; with other examples in [We2].

[Ka] has shown for any $p \geq 3$ there are closed orientable surfaces of genus p immersed in $\mathbf{R}^3$ with constant mean curvature.

CHAPTER IV. REDUCTION TECHNIQUES

INTRODUCTION

Our reduction theory is based on the commutative diagram

(*)
$$\begin{array}{ccc} M & \xrightarrow{\varphi} & N \\ \rho \downarrow & & \sigma \downarrow \\ P & \xrightarrow{\bar{\varphi}} & Q \end{array}$$

where ρ, σ are Riemannian submersions with basic tension fields; or, more generally, isoparametric maps. The main idea is to characterize harmonicity or minimality of φ in terms of extremal properties of $\bar{\varphi}$. Many illustrations and applications are given in subsequent Chapters.

We begin by deriving the main properties of Riemannian submersions $\rho : M \to P$; in particular, *ρ is a harmonic map iff its fibres are minimal submanifolds.*

In Section 2 harmonic Riemannian submersions are generalized to *harmonic morphisms*
$\varphi : M \to N$. The special case $N = S^1$ is illustrated in full detail. In Chapter X an interesting family of harmonic morphisms from 3–dimensional ellipsoids into S^2 will be described, in the context of the equivariant Hopf construction.

In Section 3 we introduce the notions of *isoparametric maps and functions*. (These can be thought of as stratified Riemannian submersions with basic tension fields).

If a compact connected Lie group G acts on a manifold N by isometries, the projection $\sigma : N \to N/G$ is an isoparametric map whose leaves are the orbits of G. An associated diagram (*) is the basic setting for the study of minimal (and constant mean curvature) G–invariant submanifolds of N. That is illustrated in Sections (4.8–11), which can serve as a model for all the constructions in Chapters V, VI, VII. More specifically, we derive from Corollary (4.11) the O.D.E. characterizing minimality of 1–cohomogeneous G–invariant submanifolds.

Section 4 is the core of the Chapter. In it we establish Reduction Theorems (4.2), (4.5), (4.13). Clearly these provide methods reaching far beyond the case of group equivariance – as we shall see in Chapters IX, X in particular.

1. RIEMANNIAN SUBMERSIONS

(1.1) Let M and P be Riemannian manifolds and $\pi : M \to P$ a submersion (i.e., each differential $d\pi(x) : T_x(M) \to T_{\pi(x)}(P)$ is surjective); then there is a canonical splitting of the tangent bundle

$$T(M) = T^H(M) \oplus T^V(M)$$

where $T^V(M) = \operatorname{Ker} d\pi$ and $T^H(M)$ is its orthogonal complement.

(1.2) Say that π is a *Riemannian submersion* if each restriction $d\pi(x) | T_x^H(M) \to T_{\pi(x)}(P)$ is an isometry.

(1.3) Let $\pi : (M, g) \to (P, h)$ be a Riemannian submersion. A vector field $X \in \mathcal{C}(T(M))$ is *basic* if X is horizontal and there is a vector field $X_* \in \mathcal{C}(T(P))$ such that $d\pi(x) X(x) = X_*(\pi(x))$ for all $x \in M$. Horizontal lifts of vector fields on P are basic; indeed, the map $X \to X_*$ is a bijection of the basic fields of M onto the vector fields of P; and

$$g(x)(X, Y) = h(\pi(x))(X_*, Y_*)$$

for all basic vector fields X, Y.

Letting subscripts H and V indicate the projections of $T(M)$ onto $T^H(M)$ and $T^V(M)$, we note the

(1.4) **Lemma.** *If X, Y are basic vector fields, then*

(a) $d\pi[X, Y] = [X_*, Y_*]$; so $([X, Y]_H)_* = [X_*, Y_*]$

(b) $\left((\nabla^M_X Y)_H\right)_* = \nabla^P_{X_*} Y_*$.

Proof. The first assertion is immediate. To verify (b) we use the characterization

$$2g(\nabla_X Y, Z) = X g(Y, Z) + Y g(Z, X) - Z g(X, Y)$$
$$- g(X, [Y, Z]) + g(Y, [Z, X]) + g(Z, [X, Y])$$

to see that for every $x \in M$,

$$2g(x)\left(\nabla^M_X Y(x), Z(x)\right) = 2h(\pi(x))(\nabla^P_{X_*} Y_*(\pi(x)), Z_*(\pi(x))) . //$$

(1.5) **Lemma.** *Let $\beta(\pi) = \nabla d\pi$ denote the second fundamental form of the Riemannian submersion $\pi : M \to P$. Then*

(a) $\beta(\pi)|T^H(M) \times T^H(M) = 0$

(b) $\beta(\pi)|T^V(M) \times T^V(M) = 0$ iff the fibres are totally geodesic submanifolds.

(c) $\beta(\pi)|T^H(M) \times T^V(M) = 0 = \beta(\pi)|T^V(M) \times T^H(M)$ iff the horizontal distribution is integrable.

Proof. For any $X, Y \in \mathcal{C}(T(M))$ we have

(1.6) $$\beta(\pi)(X,Y) = \nabla^P_{d\pi(X)} d\pi(Y) - d\pi(\nabla^M_X Y)$$

as in I (1.5). To verify (a) at any point $x \in M$, take $X, Y \in T^H_x(M)$ and extend them to be basic on M. Then

$$\beta(\pi)(X,Y) = \nabla^P_{X_*} Y_* - d\pi(\nabla^M_X Y) = 0$$

by (b) of Lemma (1.4).

To prove (b), let $X, Y \in T^V_x(M)$, and denote by $i_x : \pi^{-1}(\pi(x)) \to M$ the inclusion map of the fibre through x; with induced metric on the fibre, i_x is a Riemannian immersion. Then

$$0 = \beta(\pi \circ i_x) = d\pi(\beta(i_x)) + \beta(\pi)(di_x, di_x)$$

so at x we have

$$\beta(\pi)(X,Y) = -d\pi(\beta(i_x)(X,Y)).$$

Because $\beta(i_x)(X,Y)$ is horizontal by I (2.2) and $d\pi$ is an isomorphism on horizontal vectors, the right member vanishes iff $\beta(i_x) \equiv 0$. We conclude that $\beta(\pi)(X,Y) = 0$ on M iff the fibres are totally geodesic.

We proceed to the proof of (c). Take any $X, Y, Z \in \mathcal{C}(T(M))$. From (1.6) we obtain

(1.7) $$\beta(\pi)(X_V, Y_H) = -d\pi\left((\nabla_{Y_H} X_V)_H\right).$$

Now

$$\begin{aligned}
\langle (\nabla_{Y_H} X_V)_H, Z \rangle &= \langle \nabla_{Y_H} X_V, Z_H \rangle \\
&= -\langle X_V, \nabla_{Y_H} Z_H \rangle \\
&= -\langle X, (\nabla_{Y_H} Z_H)_V \rangle.
\end{aligned}$$

Consequently,

(1.8) $$(\nabla_{Y_H} X_V)_H = 0 \quad \text{for all} \quad X, Y \quad \text{iff}$$

$$(\nabla_{Y_H} Z_H)_V = 0 \quad \text{for all} \quad Y, Z.$$

Then (1.7) and (1.8) together show that

(1.9) $\qquad \beta(\pi)|T^V(M) \times T^H(M) = 0 \quad \text{iff}$

$\qquad \nabla_{Y_H} Z_H \quad \text{is horizontal for all} \quad Y, Z.$

But $\nabla_{Y_H} Z_H - \nabla_{Z_H} Y_H = [Y_H, Z_H]$, so that (1.9) is true iff the bracket of horizontal fields is horizontal. //

An immediate consequence is

(1.10) Corollary. *If $\pi : M \to P$ is a Riemannian submersion with positive fibre dimension, then the following assertions are equivalent:*

(a) *π is a totally geodesic map;*

(b) *the fibres of π are totally geodesic submanifolds, and the horizontal distribution is integrable.*

If those conditions are met and M is simply connected, then M is a Riemannian product with projection π.

(1.11) Let m, p denote the dimensions of M, P respectively. At a point $x \in M$ we can always choose an orthonormal base $(e_i)_{1 \le i \le m}$ for $T_x(M)$ with $(e_i)_{1 \le i \le p}$ spanning $T_x^H(M)$ and $(e_j)_{p+1 \le j \le m}$ spanning $T_x^V(M)$. We shall write $\tilde{e}_i = d\pi(e_i)$, $1 \le i \le p$ and say that the frame (e_i) is *adapted* to π.

(1.12) Proposition. *A Riemannian submersion $\pi : M \to P$ has constant energy density $e(\pi) = p/2$. It is a harmonic map iff its fibres are minimal submanifolds.*

Proof. The first assertion follows immediately by taking an orthonormal base at a point, the first p vectors of which are horizontal.

Consider the fibre $F_y = \pi^{-1}(y)$ over $y \in P$, with induced Riemannian metric; and let $i_y : F_y \to M$ be the inclusion map. Then

$$0 = \tau(\pi \circ i_y) = d\pi(\tau(i_y)) + \text{Trace } \beta(\pi)(di_y, di_y).$$

Now $\tau(i_y)$ is horizontal and $d\pi$ is an isometry on horizontal vectors; therefore as in the proof of Lemma (1.5) it follows that $\tau(i_y) = 0$ iff Trace $\beta(\pi)(di_y, di_y) = 0$. But by Lemma (1.5) that is equivalent to $\tau(\pi) = 0$, since

$$\tau(\pi) = \sum_{i=1}^{p} \beta(\pi)(e_i, e_i) + \sum_{j=p+1}^{m} \beta(\pi)(e_j, e_j)$$

for an adapted frame (e_i). //

(1.13) **Proposition.** *Let $\pi : M \to P$ be a Riemannian submersion, and $\psi : P \to Q$ a map. Then $e(\psi \circ \pi) = e(\psi)$; and*

$$\tau(\psi \circ \pi) = d\psi(\tau(\pi)) + \tau(\psi) \circ \pi .$$

In particular, suppose that π is harmonic; then ψ is harmonic iff $\psi \circ \pi$ is.

Proof. At a point $x \in M$ take an adapted frame (e_i) as in (1.11). Then at x

$$2e(\psi \circ \pi) = \sum_{i=1}^{m} |d\psi \circ d\pi(e_i)|^2 = \sum_{i=1}^{p} |d\psi(\tilde{e}_i)|^2 = 2e(\psi) .$$

Also, by Proposition I (1.14),

$$\tau(\psi \circ \pi) = d\psi \circ \tau(\pi) + \sum_{i=1}^{m} \beta(\psi)(d\pi(e_i), d\pi(e_i)) ;$$

and that sum equals $\sum_{i=1}^{p} \beta(\psi)(\tilde{e}_i, \tilde{e}_i) = \tau(\psi) \circ \pi$. //

2. HARMONIC MORPHISMS AND MAPS INTO A CIRCLE

(2.1) Say that a map $\varphi : M \to N$ is *horizontally conformal* if for any point $x \in M$ at which $d\varphi(x) \neq 0$ its restriction to the orthogonal complement of $\text{Ker}(d\varphi(x))$ in $T_x(M)$ is conformal and surjective. A *harmonic morphism* is a harmonic horizontally conformal map.

The factor of conformality of a harmonic morphism is a function $\lambda : M \to \mathbb{R}(\geq 0)$ whose square λ^2 is smooth and satisfies

(2.2) $$\Delta^M(f \circ \varphi) = \lambda^2(\Delta^N f) \circ \varphi$$

for all C^2 functions $f : N \to \mathbb{R}$.

Note that *a harmonic morphism has $\lambda \equiv 1$ iff it is a harmonic Riemannian submersion*.

Proposition (1.13) admits the following immediate generalization:

(2.3) **Proposition.** *Let $\varphi : M \to N$ be a surjective harmonic morphism and $\psi : N \to P$ a map. Then ψ is harmonic iff $\psi \circ \varphi$ is harmonic. In fact*

$$e(\psi \circ \varphi) = \frac{2}{n} e(\psi) e(\varphi) \quad \text{and} \quad \tau(\psi \circ \varphi) = \frac{2e(\varphi)}{n} \tau(\psi) \circ \varphi$$

where $n = \dim N$.

Moreover, if ψ is a harmonic morphism, then so is $\psi \circ \varphi$.

(2.4) The next result can be proved by using elementary properties of the stress–energy tensor $S_\varphi = e(\varphi)g - \varphi^*h$.

Theorem. *Let $\varphi : (M, g) \to (N, h)$ be a harmonic morphism which is a submersion. Then*

(a) *if $n = 2$, the fibres are minimal submanifolds;*

(b) *if $n \geq 3$, the following are equivalent:*

(i) *the fibres are minimal;*

(ii) *$\nabla e(\varphi)$ is everywhere vertical;*

(iii) *the horizontal distribution $T^H(M)$ has mean curvature $\nabla e(\varphi)/2e(\varphi)$.*

(2.5) Clearly harmonic maps $M \to S^1$ are harmonic morphisms. They can be determined explicitly as follows:

Let M be compact and oriented. The differential $d\omega$ of a 1–form $\omega \in C(T^*(M))$ is the 2–form characterized by

$$d\omega(X_1, X_2) = (\nabla_{X_1}\omega)X_2 - (\nabla_{X_2}\omega)X_1$$

for all $X_1, X_2 \in C(T(M))$. Its codifferential $d^*\omega$ is the function

(2.6) $$d^*\omega = -g^{st}(\nabla_{e_t}\omega)(e_s) = -\text{Trace } \nabla\omega,$$

where $(e_i)_{1 \leq i \leq m}$ is a local frame field for $T(M)$.

Since M is compact, ω is a harmonic form iff $d\omega = 0 = d^*\omega$. Clearly $\nabla \omega = 0$ implies that ω is harmonic.

If ω is a closed 1–form with integral periods (i.e., $d\omega = 0$ and $\int_Z \omega \in \mathbf{Z}$ for every integral 1–cycle Z on M) and $a \in M$, then the integral over an oriented path γ_{ax} from a to x

(2.7) $$\varphi(x) = \int_{\gamma_{ax}} \omega$$

is well defined and independent of the choice of path, modulo $\mathbf{Z}$. Thus (2.7) defines a map $\varphi : M \to S^1$, and $d\varphi = \omega$.

(2.8) **Proposition.** *Suppose that ω is a harmonic 1-form on M with integral periods. Then*

(a) *The map $\varphi : M \to S^1$ defined by (2.7) is a harmonic morphism.*

(b) *If ω is parallel (i.e. $\nabla \omega = 0$), then φ is totally geodesic.*

(c) *If $\omega \neq 0$ at every point, then φ is a harmonic Riemannian submersion.*

Proof. Assertions (a) and (b) are immediate. For that of (c), let $u \in \mathcal{C}(T(M))$ be the contravariant representative of ω : $< u(x), v > = \omega(x)(v)$ for every $v \in T_x(M)$. Then every $u(x)$ is orthogonal to $T_x^V(M) = \text{Ker } d\varphi(x)$; and $u(x)$ spans $T_x^H(M)$. Also, $|u(x)|^2 = |\omega(x) \cdot u(x)|^2 = |d\varphi(x) \cdot u(x)|^2$, so φ is a Riemannian submersion.
//

(2.9) Conversely, given a harmonic map $\varphi : M \to S^1$, its differential $d\varphi = \omega$ is clearly a harmonic 1-form on M with integral periods; it can be identified with $\varphi^*\vartheta$, where ϑ is the harmonic generating 1-form of S^1 (= the differential of angle).

(2.10) The set $[M, S^1]$ of homotopy classes of maps $M \to S^1$ forms an abelian group, with operation induced from pointwise addition of maps. There is a natural homomorphism

$$i : [M, S^1] \to H^1(M)$$

into the 1-dimensional cohomology group of M with integral coefficients, defined as follows: for any map $\varphi \in [\varphi] \in [M, S^1]$, let

$$i[\varphi] = \varphi^*\vartheta \ .$$

It is an elementary matter to show that i is well defined; and standard that i is actually an *isomorphism* of $[M, S^1]$ onto $H^1(M)$.

Hodge's Theorem insures that there is a unique harmonic 1-form with integral periods in every class in $H^1(M)$. From the Proposition (2.8), having chosen a base point $a \in M$ we obtain

(2.11) **Corollary.** *Every homotopy class in $[M, S^1]$ contains a harmonic map, which is also a harmonic morphism.*

(2.12) **Remark.** From the covering homotopy theorem we know that a map φ :

$M \to S^1$ is null homotopic iff it admits a lift ψ

$$\begin{array}{ccc} & & \mathbb{R} \\ & \psi \nearrow & \pi \downarrow \\ M & \xrightarrow{\varphi} & S^1 \end{array}$$

where now $\pi(x) = e^{2\pi i x}$. Note that φ is harmonic iff ψ is. Since M is compact, we conclude that a harmonic map $\varphi : M \to S^1$ is null homotopic iff it is constant.

(2.13) **Example.** Let M be a compact Riemann surface. Then

a) if genus $M = 0$, then any harmonic map $\varphi : M \to S^1$ is constant;

b) if genus $M = 1$, then there is a harmonic Riemannian submersion $M \to S^1$;

c) if genus $M \geq 2$, there are harmonic morphisms $M \to S^1$ but not harmonic Riemannian submersions.

An immediate application of (2.12) is

(2.14) **Proposition.** *If $\psi : M \to N$ is a harmonic Riemannian fibration between compact oriented manifolds, then the induced homomorphism on cohomology ψ^* : $H^1(N) \to H^1(M)$ is injective.*

3. ISOPARAMETRIC MAPS

(3.1) Let (M, g) be a Riemannian manifold. A map $\pi : M \to P$ to a smooth manifold P is *transnormal* if there is a section $A \in \mathcal{C}(\odot^2 T(P))$ such that

(3.2) $$d\pi(\tilde{g}) = A \circ \pi ,$$

where $\tilde{g}$ is the contravariant representation of g. In charts (3.2) takes the form

$$g^{ij}(x)\pi_i^\alpha(x)\pi_j^\beta(x) = A^{\alpha\beta}(\pi(x)) .$$

We observe that *the rank of π is constant* on the fibres $\pi^{-1}(z)$. The fibres on which π has maximal rank are called *regular*; their union is an open subset $M^0 \subset M$.

(3.3) **Proposition.** *If $\pi : M \to P$ is transnormal then $P^0 = \pi(M^0)$ is a submanifold whose dimension is the maximal rank of π. It has a canonical metric k^0 with respect to which $\pi^0 : (M^0, g) \to (P^0, k^0)$ is a Riemannian submersion.*

Here $\pi^0 = \pi|_{M^0}$; and k^0 measures distances between the fibres of π^0.

Proof. We observe that π^0 is a submersion onto P^0; that is because for any $x \in M^0$ the differential $d\pi^0(x)$ maps $T_x(M)$ onto a p–dimensional subspace $T_z(P^0)$ of $T_z(P)$, where $z = \pi(x)$. If $T_x^V(M^0) = \text{Ker } d\pi^0(x)$ and $T_x^H(M^0)$ its orthogonal complement, we define the metric k_z^0 on $T_z(P^0)$ by declaring $d\pi^0(x) : T_x^H(M^0) \to T_z(P^0)$ to be an isometry. Now (3.2) insures that k_z^0 is independent of $x \in \pi^{-1}(z)$. //

(3.4) A map $\pi : (M, g) \to (P, k)$ between Riemannian manifolds is *isoparametric* if

(i) π is transnormal; and

(ii) there is a section $B \in C(T(P))$ such that

$$\tau(\pi) = B \circ \pi .$$

(3.5) **Example.** If $P = \mathbb{R}$, then $d\pi(\tilde{g}) = |d\pi|^2$ and $\tau(\pi) = -\Delta\pi$. Therefore a function $\pi : (M, g) \to \mathbb{R}$ is isoparametric as a map iff it is isoparametric as a function (in Cartan's sense [C1,2]. See also [Ba]).

In this case, the geometric meaning of conditions (3.4) is that the level hypersurfaces of π are parallel and have constant principal curvatures.

In Chapter VIII we shall illustrate how isoparametric functions on spheres are associated with harmonic polynomial maps.

(3.6) **Example.** If $P = \mathbb{R}^r$, we recover Wang's definition [Wa]: $\pi : (M, g) \to \mathbb{R}^r$ is isoparametric iff there are functions $(a^{\alpha\beta})$, (b^γ) such that

(3.7) $< d\pi^\alpha, d\pi^\beta > = a^{\alpha\beta}(\pi^1, \ldots, \pi^r)$ and $\Delta\pi^\gamma = b^\gamma(\pi^1, \ldots, \pi^r)$.

(3.8) Let $\pi : (M, g) \to (P, k)$ be a Riemannian submersion. Say that π has *basic tension field* if $\tau(\pi)(x) = \tau(\pi)(x')$ whenever $\pi(x) = \pi(x')$.

(3.9) **Proposition.** *A Riemannian submersion $\pi : (M, g) \to (P, k)$ with basic tension field is an isoparametric map.*

Proof. Let $A = \tilde{k}$, the contravariant representative of k; then $d\pi(\tilde{g}) = A \circ \pi$; i.e., π is transnormal. The condition (3.8) is equivalent to the existence of a vector field $B \in C(T(P))$ such that $\tau(\pi) = B \circ \pi$. //

(3.10) Proposition. *If $\pi : (M, g) \to (P, k)$ is isoparametric, then*

(a) *the mean curvature vector of the regular fibres of π is a basic vector field on M^0; and*

(b) *the Riemannian submersion $\pi^0 : (M^0, g) \to (P^0, k^0)$ has basic tension field.*

Proof. As in the proof of Proposition (1.12),
$$0 = \tau(\pi^0 \circ i_z)(x) = d\pi^0(i_z(x)) \cdot \tau(i_z)(x) + \tau(\pi^0)(i_z(x)) ;$$
we see that $\tau(\pi^0)$ is basic iff $\tau(i_z)$ is basic. The proposition now follows by application of (ii) in (3.4).

As in (1.13) we have the

(3.11) Proposition. *Let π be a Riemannian submersion:*

$$(M, g)$$
$$\pi \downarrow \quad \searrow \varphi$$
$$(P, k) \xrightarrow{\psi} (N, h) .$$

If ψ is a transnormal (resp., isoparametric) map, then so is $\psi \circ \pi = \varphi$.

(3.12) Proposition. *Let $j : (N, h) \to (P, k)$ be an isometric embedding onto a closed submanifold. If $\varphi : (M, g) \to (N, h)$ is transnormal (resp., isoparametric), then so is the composition $\Phi = j \circ \varphi$:*

$$(M, g)$$
$$\varphi \downarrow \quad \searrow \Phi$$
$$(N, h) \xrightarrow{j} (P, k) .$$

And conversely.

Proof. Given $A \in C(\odot^2 T(N))$ defining the transnormality of φ, choose any extension $\tilde{A} \in C(\odot^2 T(P))$; then $dj(y)A(y) = \tilde{A}(y)$ for $y \in N$, and
$$d\Phi(\tilde{g})(x) = dj(\varphi(x)) \circ d\varphi(x)\tilde{g}(x)$$
$$= dj(\varphi(x))A(\varphi(x)) = \tilde{A} \circ \Phi(x) .$$

Similarly, if B is as in (ii) of (3.4) with respect to φ and we extend B arbitrarily to $\tilde{B} \in C(T(P))$, then
$$\tau(\Phi) = dj \cdot \tau(\varphi) + \text{Trace } \nabla dj(d\varphi, d\varphi)$$
$$= \tilde{B} \circ \varphi + g^{ij} (\nabla dj)_{\alpha\beta} \varphi_i^\alpha \varphi_j^\beta$$
$$= (\tilde{B} + \nabla dj \tilde{A}) \circ \varphi .$$

Analogously for the converse. //

In the notation of (3.12) we have

(3.13) **Corollary.** *If $(P, k) = \mathbb{R}^r$, then φ is transnormal (resp., isoparametric) iff Φ satisfies (3.7) with π replaced by Φ.*

(3.14) **Homogeneous isoparametric maps.** Suppose that a compact connected Lie group G acts on a complete Riemannian manifold N by isometries. Then there exists an isoparametric map $\pi : N \to \mathbb{R}^s$ (for s sufficiently large) such that the fibres of π are precisely the orbits of G. This is due to a theorem of G. Schwarz [Sc]: *Every smooth G-invariant function (in particular, $< d\pi^\alpha, d\pi^\beta >, \Delta \pi^\gamma)$ is a function of a suitable finite set of G-invariant functions $\pi^1, \ldots, \pi^s$.* Thus conditions (3.7) are satisfied, and the projection $\pi : N \to N/G$ is an isoparametric map by Corollary (3.13).

In order to apply this to specific examples, it is convenient to proceed as follows:

(3.15) For any point $x \in N$, let $G(x)$ denote the orbit through x, and $G_x = \{g \in G : g \cdot x = x\}$ the isotropy group. Thus $G(x) = G/G_x$.

Say that two orbits $G(x)$ and $G(x')$ have the *same type* if G_x and $G_{x'}$ are conjugate in G.

Then the orbit types of N are the conjugacy classes of the isotropy groups $\{G_x : x \in N\}$. We partially order these as follows:

$$(H) \geq (K) \quad \text{iff } K \supset gHg^{-1} \quad \text{for some} \quad g \in G.$$

We will use the following two results on transformation groups [MSY]:

Theorem (*Lower semi-continuity of orbit types*). *In a sufficiently small G-invariant neighbourhood U of an orbit $G(x)$ there is only a finite number of orbit types, and for each $y \in U$*

$$(G_y) \geq (G_x).$$

Theorem (*Principal orbit type*). *There is a unique orbit type (H) such that $(H) \geq (K)$ for all orbit types (K) of the action. Moreover, the union of all orbits of type (H), namely $N^* = \{x \in N : G_x \in (H)\}$, is an open, dense submanifold of N.*

We call (H) the *principal orbit type*; and $H \in (H)$ a principal isotropy subgroup. If (H') is not principal, but $\dim H = \dim H'$, we say that H' is an *exceptional orbit*. In this case G/H is a finite covering of G/H'.

(3.16) Now we consider the homogeneous isoparametric map $\pi : N \to N/G$. Clearly $N^0 = N^*$ and so we have a canonical metric h on N^*/G which makes $\pi = \pi|_{N^*} : N^* \to N^*/G$ a Riemannian submersion.

(3.17) Now we introduce the volume function $V : N/G \to \mathbf{R}$:

$$V(z) = \begin{cases} \text{Vol } (\pi^{-1}(z)) & \text{if } \pi^{-1}(z) \text{ is a principal orbit} \\ m.\text{Vol } (\pi^{-1}(z)) & \text{if } \pi^{-1}(z) \text{ is an exceptional orbit} \\ 0 & \text{otherwise} \end{cases}$$

where $m = \sharp(H'/H)$ for an appropriate $H' \in (G_y), y \in \pi^{-1}(z)$. V is continuous on N/G and differentiable on N^*/G.

(3.18) Let $P \subset N^*/G$ be a submanifold of dimension k. We say that $\pi^{-1}(P) \subset N^*$ is a G-invariant submanifold of *cohomogeneity k*.

4. REDUCTION TECHNIQUES

(4.1) Let $\rho : M \to P, \sigma : N \to Q$ be Riemannian submersions. We say that a map $\varphi : M \to N$ is (ρ, σ)-*equivariant* if $\rho(x) = \rho(x')$ implies $\sigma(\varphi(x)) = \sigma(\varphi(x'))$. Equivalently, φ determines a map $\bar\varphi$ such that $\sigma \circ \varphi = \bar\varphi \circ \rho$:

$$\begin{array}{ccc} M & \xrightarrow{\varphi} & N \\ \rho\downarrow & & \downarrow\sigma \\ P & \xrightarrow{\bar\varphi} & Q \end{array}$$

Similarly, we say that a vector field w along φ is (ρ, σ)-equivariant if $d\sigma(w(x)) = d\sigma(w(x'))$ whenever $\rho(x) = \rho(x')$.

Also, we say that φ is *horizontal* if $d\varphi(T^H(M)) \subseteq T^H(N)$.

(4.2) **Theorem.** *Let φ be a (ρ, σ)-equivariant map. Assume that*

(a) *ρ has basic tension field;*

(b) *$\rho(x) = \rho(x')$ implies Trace $\varphi^*(\nabla d\sigma)(x) =$ Trace $\varphi^*(\nabla d\sigma)(x')$.*

Then φ is harmonic iff it is stationary with respect to (ρ, σ)-equivariant variations (i.e., variations through (ρ, σ)-equivariant maps).

Proof. First we observe that there is a natural bijection between (ρ, σ)-equivariant vector fields and (ρ, σ)-equivariant variations. Moreover, we know from I (1.11) that $\tau(\varphi)$ determines a direction along which the energy decreases most rapidly; thus it suffices to show that under our assumptions $\tau(\varphi)$ is (ρ, σ)-equivariant.

We apply the composition law to $\sigma \circ \varphi = \tilde{\varphi} \circ \rho$ to obtain

(4.3) $\quad d\sigma(\tau(\varphi)) = d\tilde{\varphi}(\tau(\rho)) + \text{Trace } \nabla d\tilde{\varphi}(d\rho, d\rho) - \text{Trace } \varphi^*(\nabla d\sigma)$

(Here we have used the representation

$$\text{Trace } \varphi^*(\nabla d\sigma) = \sum_{i=1}^{m} \nabla d\sigma(d\varphi(e_i), d\varphi(e_i))$$

where e_i is a local orthonormal frame for $T(M)$.)

Since ρ is a Riemannian submersion we have (as in the proof of (1.13))

(4.4) $\quad\quad\quad\quad \text{Trace } \nabla d\tilde{\varphi}(d\rho, d\rho) = \tau(\tilde{\varphi}) \circ \rho$.

Using (4.2), (a) and (b) in (4.3) we conclude immediately that $\tau(\varphi)$ is (ρ, σ)-equivariant. //

Theorem (4.2) provides very general conditions which guarantee reduction; however, it is often difficult to check whether (b) holds or not. The following is useful in the study of G-invariant minimal submanifolds, as shown in (4.8) below.

(4.5) **Theorem.** *The conclusion in Theorem (4.2) holds if (b) is replaced by these three conditions*:

(i) σ *has basic tension field*;

(ii) φ *is horizontal*;

(iii) $\varphi|_{\rho^{-1}(z)}$ *is a Riemannian submersion for all $z \in P$, with respect to the induced metrics*.

Proof. Let $(e_i)_{1 \le i \le m}$ be a frame adapted to ρ. By using (ii), (iii) and Lemma (1.5) we compute

$$\text{Trace } \varphi^*(\nabla d\sigma) = \sum_{i=1}^{m} \nabla d\sigma(d\varphi(e_i), d\varphi(e_i))$$
$$= \tau(\sigma) \circ \varphi .$$

Thus (4.3) becomes

(4.6) $\quad\quad\quad d\sigma(\tau(\varphi)) = d\tilde{\varphi}(\tau(\rho)) + \tau(\tilde{\varphi}) \circ \rho - \tau(\sigma) \circ \varphi$.

Now, since ρ, σ have basic tension fields, it is clear that $\tau(\varphi)$ is (ρ, σ)-equivariant. //

Similarly, we have the

(4.7) **Proposition.** *The conclusion in Theorem (4.2) holds if (b) is replaced by*

(i) σ *has totally geodesic fibres;*

(ii) φ *is horizontal.*

Moreover, if ρ is harmonic, then $\bar\varphi$ is harmonic iff $\tau(\varphi)$ is vertical.

As in the proof of Theorem (4.5)

$$\tau(\sigma) \circ \varphi = \text{Trace } \varphi^*(\nabla d\sigma) = 0,$$

so (4.6) reduces to

$$d\sigma(\tau(\varphi)) = d\bar\varphi(\tau(\rho)) + \tau(\bar\varphi) \circ \rho.$$

Now (4.7) follows easily. //

(4.8) **Application.** We consider G-invariant submanifolds of N of cohomogeneity $k \geq 1$, using (3.15–18). Form the pull-back construction

$$\begin{array}{ccc} \bar\varphi^{-1}(N^*) = & M & \xrightarrow{\varphi} & N^* \\ & \rho\downarrow & & \downarrow\sigma \\ & P & \xrightarrow{\bar\varphi} & N^*/G \end{array}$$

where $k = \dim P$ and M is the total space of the induced bundle; i.e., $M = \{(z, y) \in P \times N : \bar\varphi(z) = \sigma(y)\}$; and $\rho(z, y) = z$.

On N/G we take the metric h_k given by

$$h_k = V^{2/k} h$$

where V is the volume function (3.17) and h is the metric described in (3.16).

By construction the volume of M in N equals the volume of P in $(N/G, h_k)$.

Moreover, the hypotheses of Theorem (4.5) are clearly fulfilled. Together these establish the

(4.9) **Corollary.** *φ is a minimal immersion in N iff $\bar\varphi$ is a minimal immersion in $(N/G, h_k)$.*

In particular, the study of G-invariant minimal (and constant mean curvature) submanifolds of cohomogeneity 1 reduces to the study of *geodesic curves* in $(N/G, h_1)$.

(4.10) Let γ be a curve in N/G parametrized by arc length, and assume that the principal orbits have codimension 2 in N; i.e., N/G is 2-dimensional. We denote by $\tau_1(\gamma)$ the tension field of γ with respect to h_1. And by $\tau(\gamma)$ and ν the tension field and unit normal to γ with respect to h. Since $h_1 (= V^2 h)$ is a *conformal* deformation of h, it is easy to compute

$$\tau_1(\gamma) = \tau(\gamma) - \left[\frac{d}{d\nu} \log V\right] \nu.$$

As a consequence, we obtain

(4.11) **Corollary.** *A G-invariant hypersurface $M \subset N$ of cohomogeneity 1 has constant mean curvature H iff*

$$(n-1)H = k(\gamma) - \frac{d}{d\nu} \log V$$

where $k(\gamma)$ is the geodesic curvature of γ (i.e., $\tau(\gamma) = k(\gamma)\nu$).

(4.12) **Remark.** The case of cohomogeneity $k = 0$ (i.e., P is a point) is the study of the orbits of G in N which have extremal properties; that special case is illustrated in Section 1 of Chapter V.

The following is the main reduction tool for Chapters IX, X:

(4.13) **Theorem.** *Let φ be a (ρ, σ)-equivariant horizontal map. Assume that*

(i) *ρ has basic tension field;*

(ii) *All restrictions $\varphi_{z,w} : \rho^{-1}(z) \to \sigma^{-1}(w)$ (with $w = \bar{\varphi}(z)$) are harmonic (with respect to the induced metrics).*

Then φ is harmonic iff

(4.14) $$d\bar{\varphi}(\tau(\rho)) + \tau(\bar{\varphi}) \circ \rho - \text{Trace } \varphi^*(\nabla d\sigma) = 0.$$

Proof (which has benefited from a comment by L. Lemaire). This follows immediately from (4.3) and (4.4) provided that we show that $[\tau(\varphi)]^V = 0$. For $w = \bar{\varphi}(z)$

we consider

$$\begin{array}{ccc} M & \xrightarrow{\varphi} & N \\ i_z \uparrow & & \uparrow j_w \\ \rho^{-1}(z) & \xrightarrow{\varphi_{z,w}} & \sigma^{-1}(w) . \end{array}$$

Because φ is horizontal, $\tau(\varphi)$ and $\sum_{j=p+1}^{m} \nabla d\varphi(e_i, e_i)$ have the same vertical components.

Indeed, if (e_i) is an adapted orthonormal frame field as in (1.11) then $\nabla_{e_i} e_i$ is horizontal ($1 \leq i \leq p$): given any $X \in C(T^V(M))$, we have

$$0 = \frac{1}{2}X. <e_i, e_i> = <\nabla_X e_i, e_i> = <\nabla_{e_i} X + [X, e_i], e_i> = <\nabla_{e_i} X, e_i>$$

because $[X, e_i]$ is vertical by Lemma (1.4a); then

$$0 = <\nabla_{e_i} X, e_i> = e_i. <X, e_i> - <X, \nabla_{e_i} e_i> = - <X, \nabla_{e_i} e_i> .$$

Similarly, $\nabla^{\varphi}_{e_i} (d\varphi.e_i)$ is horizontal ($1 \leq i \leq p$), in the notation of (1.5); and therefore so is

$$\nabla d\varphi(e_i, e_i) = \nabla_{e_i}(d\varphi.e_i) = \nabla^{\varphi}_{e_i}(d\varphi.e_i) - d\varphi(\nabla_{e_i} e_i) \quad \text{for} \quad 1 \leq i \leq p .$$

In conclusion, we have

$$[\tau(\varphi)]^V = \Big[\sum_{j=p+1}^{m} \nabla d\varphi(e_i, e_i) \Big]^V = [\tau(\varphi \circ i_z) - d\varphi(\tau(i_z))]^V = [\tau(\varphi \circ i_z)]^V$$

because $\tau(i_z)$ is horizontal. But $[\tau(\varphi \circ i_z)]^V = [\tau(j_w \circ \varphi_{z,w})]^V$, and therefore

$$[\tau(\varphi)]^V = [\tau(j_w \circ \varphi_{z,w})]^V = dj_w(\tau(\varphi_{z,w})) = 0$$

because the $\varphi_{z,w}$ are harmonic. //

(4.15) **Application.** Assume that ρ, σ are determined by isoparametric functions on M, N; thus P and Q are 1–dimensional. In Chapters IX,X we shall illustrate some favourable instances where the hypotheses of Theorem (4.13) hold. And also (b) of (4.2). In this case, (4.14) is a second order *ordinary* differential equation, with prescribed boundary conditions.

5. Notes and comments

(5.1) The geometry of Riemannian submersions was initiated by O'Neill [ON]; and carried on by Vilms [V], who proved the following general structure theorem for totally geodesic maps: *If (M,g) is complete then each totally geodesic map $\varphi : (M,g) \to (N,h)$ factors into a totally geodesic Riemannian submersion followed by a totally geodesic Riemannian immersion.*

(5.2) Proposition (1.12) is due to Eells–Sampson [ES]. And Proposition (1.13) to Smith [Sm1]. That was generalized in Proposition (2.3) to harmonic morphisms, by Fuglede and Ishihara ([Fu], [I1]). Harmonic morphisms $\varphi : M \to N$ are precisely those maps which induce a morphism of the sheaf of harmonic functions on N to the sheaf of harmonic functions on M [Fu]. Theorem (2.4) is due to Baird-Eells [BE].

(5.3) The notion of isoparametric map introduced in Section 3 (specially designed for reduction theory) was proposed by Eells–Wang (Spring 1987). It is broader than the concept introduced by Carter–West [CW], which requires a further integrability condition on the horizontal distribution. In case $N = \mathbb{R}^n$, the definition of [CW] is equivalent to that of Terng [Te], except that the latter permits weaker differentiability conditions; their equivalence has been demonstrated by Wang and West [We]. All these definitions coincide with that studied by E. Cartan ([C1], [C2]) in case $N = \mathbb{R}$. A survey of the subject can be found in [Wa].

(5.4) A study of reduction theory for minimal immersions in a group–theoretic context was made by Hsiang–Lawson [HL], following an important example by Bombieri, de Giorgi, Giusti [BDG]. That was carried further in [Sm1,2] for harmonic maps.

Baird [Ba] and Pluzhnikov [Pl1] introduced the method of isoparametric functions in reduction theory. Improvements were made by Karcher–Wood [KW]. Further steps were taken in [PT].

PART 2. G–INVARIANT MINIMAL AND CONSTANT MEAN CURVATURE IMMERSIONS

CHAPTER V. FIRST EXAMPLES OF REDUCTION

INTRODUCTION

In Section 1 we work in a group theoretic context to present examples of harmonic maps and minimal immersions of cohomogeneity zero. Existence of extremal maps is reduced to finding critical points of a suitable real valued function defined on (part of) the target. In particular, we prove that every compact homogeneous space G/H endowed with a G-invariant metric, can be harmonically immersed in some Euclidean sphere S^n; and also Hsiang's theorem (1.13) that G/H can be minimally immersed in some S^n. Also, we construct a family of $SU(2)$-equivariant harmonic maps from $S^3 = SU(2)$, with a suitable left-invariant metric, to $\mathbb{C}P^1 = S^2$. Most of the harmonic maps of this family are *not* harmonic morphisms; that is of interest because all the previously known harmonic maps from S^3 with given metric to S^2 are harmonic morphisms (see Chapter X).

Section 2 illustrates examples of minimal submanifolds of cohomogeneity 1; namely, a class of $SO(n-1)$-invariant minimal hypersurfaces of S^n, found by Otsuki [Ot]; and a family of $SO(2)$-invariant minimal surfaces in S^3, discovered by Hsiang and Lawson [HL]. Much in the spirit of this monograph, existence results are obtained from the qualitative study of an O.D.E.

In Section 3 we describe Delaunay's classification of constant mean curvature surfaces of revolution in $\mathbb{R}^3$; and some natural generalizations to $SO(n-1)$-invariant hypersurfaces in $\mathbb{R}^n$. In particular, we give a simple, self-contained proof of the existence of periodic solutions. The importance of periodicity for our purposes is twofold: it guarantees that the Gauss map of the above hypersurfaces factors through a harmonic map $S^{n-2} \times S^1 \to S^{n-1}$; and it provides a tool for the proof of some analytical lemmas in Chapter VII.

1. G–EQUIVARIANT HARMONIC MAPS

(1.1) If G is a compact Lie group acting isometrically on a compact manifold M, then the tension field of the immersion $\varphi : G(x) \to M$ of an orbit is G-equivariant:

For any $g \in G$,

$$dg \cdot \tau(\varphi) = \text{Trace } \nabla(dg \circ d\varphi)$$
$$= \text{Trace } (\nabla d\varphi) \circ g = \tau(\varphi) \circ g .$$

(1.2) Let G be a group of isometries of both M and N. Then G also acts on the space $C(M, N)$ of maps of M into N, by

$$(g \cdot \varphi)(x) = g \cdot \varphi(g^{-1} \cdot x) \quad \text{for all} \quad x \in M ,$$

where $g \in G$ and $\varphi \in C(M, N)$. Then the energy functional

$$E : C(M, N) \to \mathbb{R}$$

is G–invariant: $E(g \cdot \varphi) = E(\varphi)$; i.e., E is constant on the orbits. Indeed,

$$|d(g \circ \varphi)|^2 = |d\varphi|^2 \circ g^{-1}$$

at each point of M, because each g – and hence its differential dg – is an isometry. Also, $g^* dx = (dx) \circ g^{-1}$ for the same reason.

Let $C(M, N)^G$ denote the set of fixed points of the G–action on $C(M, N)$. Thus $\varphi(g \cdot x) = g \cdot \varphi(x)$ for all $g \in G$, *the equivariant maps*. For these,

$$\tau(\varphi)(g \cdot x) = dg \cdot \tau(\varphi)(x) \quad \text{for all} \quad x \in M .$$

Furthermore, every $\varphi_t \in C(M, N)^G$, where

(1.3) $$\varphi_t(x) = \exp_{\varphi(x)}(t\tau(\varphi)(x)) ;$$

indeed, for $g \in G$,

$$\varphi_t \cdot g = \exp(dg \cdot t\tau(\varphi_0))$$
$$= g \circ \exp(t\tau(\varphi_0)) = g \cdot \varphi_t .$$

(1.4) **G–Reduction Theorem.** *Assume M compact. Then every critical point of $E|_{C(M,N)^G}$ is a critical point of E. Equivalently, $\varphi \in C(M, N)^G$ is a critical point of E iff φ is stationary with respect to equivariant variations.*

Proof. For any $\varphi_0 \in C(M, N)^G$, the above variation (1.3) satisfies

$$\left. \frac{dE(\varphi_t)}{dt} \right|_{t=0} = - \int_M |\tau(\varphi_0)|^2 \, dx .$$

Thus if $\left.\frac{dE(\varphi_t)}{dt}\right|_{t=0} = 0$ for every variation in $C(M,N)^G$, then it does in particular for $\tau(\varphi_0)$. We conclude that $|\tau(\varphi_0)|^2 \equiv 0$, so φ_0 is a critical point of E. //

(1.5) **Application.** Let $M = G/H$ be a compact Riemannian homogeneous space, endowed with a G-invariant Riemannian metric. Fix a point $x_0 \in M$, and let H be its stabilizer ($H = \{g \in G : g \cdot x_0 = x_0\}$). Clearly, if $\varphi \in C(M,N)^G$, then H is in the stabilizer of $\varphi(x_0)$.

Let $N^H = \{y \in N : b \cdot y = y \text{ for all } b \in H\}$; then we have a bijection

(1.6) $$N^H \xrightarrow{\sim} C(M,N)^G$$

given by $y \to \varphi_y$, where

(1.7) $$\varphi_y(g \cdot x_0) = g \cdot y.$$

(1.8) N^H is a closed submanifold of N; and because G acts transitively on M, we conclude that the energy density $e(\varphi_y)$ is constant. Therefore we can define the function

(1.9) $$\varepsilon : N^H \to \mathbb{R}$$

by $\varepsilon(y) = e(\varphi_y)(x_0)$.

(1.10) **Lemma.** *The critical points of $E|_{C(M,N)^G}$ correspond under (1.6) to those of ε.*

Proof. If $y \in N^H$ is a critical point of ε, let φ_t be the variation of φ_y determined by $\tau(\varphi_y)$. Then $\varphi_t \in C(M,N)^G$, so $\varphi_t = \varphi_{y(t)}$ for $y(t) = \varphi_t(x_0) \in N^H$. And

$$\left.\frac{d}{dt} E(\varphi_t)\right|_{t=0} = \text{Vol}(M) \left.\frac{d}{dt} \varepsilon(y(t))\right|_0 = 0,$$

so φ_y is a critical point of E. And conversely. //

(1.11) **Corollary.** *Let $M = G/H$, with any G-invariant metric. Assume that M and N are compact. If $\varphi_0 \in C(M,N)^G$ is a non-constant map, then φ_0 is homotopic in $C(M,N)^G$ to a harmonic non-constant $\varphi_1 \in C(M,N)^G$.*

Proof. N^H is a closed submanifold of N, so it is compact. Let $y_0 \in N^H$ correspond to φ_0 under the bijection (1.6); then there exists a point y_1 which maximizes

ε on the component of N^H containing y_0. Clearly $\varepsilon(y_1) \geq \varepsilon(y_0) > 0$. Let φ_1 correspond to y_1. //

(1.12) **Theorem.** *Let $M = G/H$ be a compact homogeneous space endowed with any G-invariant metric. Then M can be harmonically immersed into some Euclidean sphere.*

Proof. We use the notations of IV (3.15). A theorem of Mostow–Palais insures that M can be realized as an orbit of a linear G-action on S^n, for some n; i.e., there is a faithful representation of G in $O(n+1)$ and $x_0 \in S^n$ such that

$$H = \{g \in G : g \cdot x_0 = x_0\} ;$$

that guarantees that the inclusion $G(x_0) \hookrightarrow S^n$ is an embedding of M in S^n.

Let S be the component of $(S^n)^H$ which contains x_0; and let y be an ε-maximum on S. If $G(y)$ is an orbit of type (H) we are done – and the map φ_y associated with y under (1.6) is actually a harmonic *embedding* of M in S^n. Suppose now that $G(y)$ is an orbit of type $(H') \neq (H)$; since $y \in (S^n)^H$, it is obvious that $(H') \leq (H)$.

We claim that (H') must be an exceptional orbit (i.e., $\dim H' = \dim H$ and $G/H \to G/H'$ is a finite covering); otherwise, as a consequence of IV (3.17) its volume would be zero, which is impossible since $\varepsilon(y) > 0$. Thus we have a harmonic embedding $\varphi_y : G/H' \to S^n$; and composition with the covering map $G/H \to G/H'$ gives the required harmonic immersion. //

Similar techniques were first introduced for minimal immersions:

(1.13) **Theorem.** *Any compact homogeneous space G/H can be minimally immersed into some S^n.*

The proof requires the following modifications of that of (1.12): Firstly, we replace ε by the volume function V of IV (3.17). The methods of Corollary IV (4.9) give a bijection between critical points of V and G-equivariant minimal immersions. Let $(S^n)_H = \{x \in S^n : G_x \in (H)\}$; and N' the component of $(S^n)_H$ which contains x_0.

Now maximize V on $\overline{N'/G}$:

(a) If a V-maximum $y \in N'/G$, then its associated orbit is a minimal embedding of G/H into S^n.

(b) If $y \in \overline{N'/G} - N'/G$, then the theorem on lower semi–continuity of orbit types in IV (3.15) insures that its orbit type (H') satisfies $(H') \leq (H)$. We can now proceed as in (1.12). //

(1.14) Here we describe a procedure to compute explicitly the function ε: Let x_0 denote the coset H in $G/H = M$; then $T_{x_0}(M)$ is identified with $\mathcal{G}/\mathcal{H}$, where $\mathcal{G}$ and $\mathcal{H}$ are the Lie algebras of G and H respectively. There is a canonical correspondence between G-invariant Riemannian metrics on M and $ad(H)$-invariant inner products on $\mathcal{G}/\mathcal{H}$; let us fix one such, and let $(\xi_i)_{1 \leq i \leq m}$ be any set of vectors in $\mathcal{G}$ which projects to an orthonormal basis of $\mathcal{G}/\mathcal{H}$. Each vector ξ_i defines a vector field $\tilde{\xi}_i$ on M by

$$(\tilde{\xi}_i)(x) = \frac{d}{dt}\left[(\exp t\, \xi_i) \cdot x\right]\bigg|_{t=0}.$$

Similarly, if G acts on (N, h), we have vector fields on N

(1.15) $$(\hat{\xi}_i)(y) = \frac{d}{dt}\left[(\exp t\, \xi_i) \cdot y\right]\bigg|_{t=0}.$$

If $\varphi \in C(M, N)^G$, then

$$d\varphi(x)(\tilde{\xi}_i) = \hat{\xi}_i(\varphi(x)).$$

Thus

$$\varepsilon(y) = e(\varphi_y)(x_0) = \frac{1}{2}\sum_{i=1}^{m} (\varphi_y^* h)(x_0)(\tilde{\xi}_i(x_0), \tilde{\xi}_i(x_0))$$

$$= \frac{1}{2}\sum_{i=1}^{m} h(y)(\hat{\xi}_i(y), \hat{\xi}_i(y))$$

$$= \frac{1}{2}\sum_{i=1}^{m} |\hat{\xi}_i|^2.$$

In summary:

(1.16) **Proposition.** *Let G be a compact Lie group of isometries of N and H a closed subgroup of G. If $(\xi_i)_{1 \leq i \leq m}$ is an orthonormal basis in $\mathcal{G}$ for an $ad(H)$-invariant inner product on $\mathcal{G}/\mathcal{H}$, then the critical points y of*

$$\varepsilon = \frac{1}{2}\sum_{i=1}^{m} |\hat{\xi}_i|^2$$

on N^H are in bijective correspondence with the equivariant harmonic maps $\varphi_y : G/H \to N$, where G/H is given the G-invariant metric associated with the inner product on $\mathcal{G}/\mathcal{H}$.

(1.17) Application. Let S^1 act isometrically on N. The orbit of a point $y \in N$ is represented by the map $\gamma_y : S^1 \to N$ given by

$$\gamma_y(\theta) = \theta \cdot y \quad \text{for} \quad \theta \in S^1.$$

We denote by $Y \in C(TN)$ the Killing field $Y(y) = \gamma_y'(0)$. Now S^1 induces an action on $C(S^1, N)$ by $(t \cdot \gamma)(\theta) = t \cdot \gamma(\theta - t)$. And $y \to \gamma_y$ is a diffeomorphism of N onto $C(S^1, N)^{S^1}$. γ_y *is a closed geodesic iff y is a critical point of $\varepsilon : N \to \mathbf{R}$, where $\varepsilon(y) = \frac{1}{2} \|Y(y)\|^2$.*

In fact, if $\theta \in S^1$,

$$\gamma_y'(\theta) = \gamma_{\theta \cdot y}'(0) = d\theta(y) \, \gamma_y'(0) = d\theta(y) \, Y(y),$$

so $\|\gamma_y'(\theta)\|^2 = \|Y(y)\|^2$; from this we conclude

$$e(\gamma_y) = \frac{1}{2} \|Y(y)\|^2 = \varepsilon(y).$$

(1.18) Application. Let $\theta : G \to G'$ be a homomorphism carrying the subgroups $H \to H'$. That determines a G'-action on G'/H'.

The induced map $\varphi = \varphi^\theta : G/H \to G'/H'$ is harmonic iff the identity coset in G'/H' is a critical point of the function

$$\varepsilon : G'/H' \to \mathbf{R}$$

given by $bH' \to e(\varphi_b^\theta)$, where

$$\varphi_b^\theta(aH) = \theta(a)bH' \quad \text{for} \quad a \in G.$$

(1.19) Application. Our object now is to produce various equivariant harmonic maps $\varphi : SU(2) \to \mathbf{C}P^2$, with respect to left–invariant Riemannian metrics on $SU(2)$ and the Fubini–Study metric on $\mathbf{C}P^2$. Some of these factor through totally geodesic embeddings $\mathbf{C}P^1 \hookrightarrow \mathbf{C}P^2$ to give harmonic maps $(S^3, g) \to S^2$ of Hopf invariant one which are not harmonic morphisms.

In Proposition (1.16) we take $G = SU(2)$, represented as the subgroup

$$\begin{pmatrix} U & 0 \\ 0 & 1 \end{pmatrix}$$

of $SU(3)$, with $U \in SU(2)$; $H = (Id)$. And $N = \mathbb{C}P^2$; here $SU(2)$ acts on $\mathbb{C}P^2$ through its inclusion in $SU(3)$. The Lie algebra $su(2)$ of $SU(2)$ consists of the traceless skew–Hermitian 2×2 matrices. The vectors

$$\xi_1 = \frac{1}{\sqrt{2}} \begin{pmatrix} i & 0 \\ 0 & -i \end{pmatrix}, \quad \xi_2 = \frac{1}{\sqrt{2}} \begin{pmatrix} 0 & 1 \\ -1 & 0 \end{pmatrix}, \quad \xi_3 = \frac{1}{\sqrt{2}} \begin{pmatrix} 0 & i \\ i & 0 \end{pmatrix}$$

form a base, which is orthonormal for the bi–invariant metric. Denote an arbitrary element $\xi \in su(2)$ by $\xi = \Sigma\, a_i\, \xi_i = \xi(a)$; and construct a left–invariant inner product on $su(2)$ by

(1.20) $$(\xi, \xi) = a^t \cdot C^{-1} \cdot a$$

where C is a positive definite 3×3 matrix.

(1.21) **Lemma.** *If $[v] = [v_1 : v_2 : v_3]$ denote homogeneous coordinates on $\mathbb{C}P^2$, then $\varepsilon : (\mathbb{C}P^2)^H = \mathbb{C}P^2 \to \mathbb{R}$ is given by*

$$\varepsilon([v]) = \frac{1}{2} \left(\|u\|\, Tr.C - u^t C\, u \right)$$

where

$$u = \frac{1}{\|v\|^2} \begin{pmatrix} |v_1|^2 - |v_2|^2 \\ 2\, Im(\bar v_1 \cdot v_2) \\ 2\, Re(\bar v_1 \cdot v_2) \end{pmatrix}.$$

Proof. The projection $\pi : \mathbb{C}^3 - \{0\} \to \mathbb{C}P^2$ is $SU(3)$–equivariant, and (up to a constant factor) is a Riemannian submersion relative to the metric $g = dv^2/|v|^2$ on $\mathbb{C}^3 - \{0\}$.

We compute

$$\hat\xi([v]) = \frac{d}{dt} \left[(\exp t\xi) \cdot [v] \right]_{t=0}$$
$$= d\pi \left(\frac{d}{dt} \left[(\exp t\xi) \cdot v \right]_{t=0} \right) = d\pi(\xi v)_v$$

where $u_v \in T_v(\mathbb{C}^3 - \{0\})$ stands for $\frac{d}{dt}[v + tu]_{t=0}$.

Since π is a Riemannian submersion,

$$|\hat\xi|^2([v]) = g\left((\xi v)_v^H, (\xi v)_v^H\right)$$
$$= \left\{ \|\xi v\|^2 - \frac{|<\xi v, v>|^2}{\|v\|^2} \right\} / \|v\|^2$$

where H denotes projection to the horizontal distribution. Then a simple computation shows that if $\xi = \xi(a)$,

$$|\hat{\xi}|^2([v]) = \frac{1}{2} \left\{ a^t \, H_{[v]} \, a \right\} / \|v\|^2$$

where

$$H_{[v]} = \|u\| 1 - uu^t .$$

Now observe that $C^{1/2}\xi_i$ is an orthonormal base for the inner product (1.20). Therefore

$$\varepsilon([v]) = \frac{1}{2} \sum_{i=1}^{3} |\widehat{C^{1/2}\xi_i}|^2 =$$

$$\frac{1}{2} Tr \left(C^{1/2} \, H_{[v]} \, C^{1/2} \right) = \frac{1}{2} Tr \left(C \, H_{[v]} \right)$$

$$= \frac{1}{2} \left(\|u\| \, Tr \, C - u^t \, C \, u \right). \, //$$

As $[v]$ varies over $\mathbb{C}P^2$, u varies over

$$\|u\|^2 = (|v_1|^2 + |v_2|^2)^2 / \|v\|^4 \leq 1 ,$$

so we look for critical points of

$$f(u) = \|u\| \, Tr \, C - u^t \, C \, u$$

on $\|u\| \leq 1$.

For the sake of simplicity, we assume that C is diagonal with eigenvalues $\lambda_1 \geq \lambda_2 \geq \lambda_3$. The critical points u of f fall into three classes:

a) $0 < \|u\| < 1$;

b) $\|u\| = 1$;

c) $u = 0$, corresponding to the constant map.

Critical points of type b) correspond to equivariant harmonic maps $\varphi : SU(2) \to \mathbb{C}P^2$ which are compositions of totally geodesic embeddings $T : \mathbb{C}P^1 \hookrightarrow \mathbb{C}P^2$ and equivariant harmonic maps $\tilde{\varphi} : SU(2) \to \mathbb{C}P^1$; i.e., $\varphi = T \circ \tilde{\varphi}$.

Such a factorization does not exist for the equivariant harmonic maps associated with critical points of type a). Now analysis of the critical points of f shows the following:

(1.22) *Let $S^3 = SU(2)$ be given a left–invariant metric as in (1.20), where C is diagonal with eigenvalues $\lambda_1 \geq \lambda_2 \geq \lambda_3$. Then there exists an equivariant harmonic map $\varphi : SU(2) \to \mathbb{C}P^2$ associated with a critical point of type a) iff $\lambda_1 > \lambda_2 + \lambda_3$.*

(1.23) As for critical points of type b), we first observe that the $SU(2)$-equivariant maps $\tilde{\varphi} : SU(2) \to \mathbb{C}P^1$ are clearly parametrized by $\mathbb{C}P^1 = S^2$. Then we have

(1.24) i) *If $\lambda_1 = \lambda_2 = \lambda_3$, all the equivariant maps $\tilde{\varphi} : SU(2) \to \mathbb{C}P^1$ are harmonic.*

ii) *If $\lambda_1 > \lambda_2 = \lambda_3$, then there are two ε–minima and an equator of ε–maxima (see figure).*

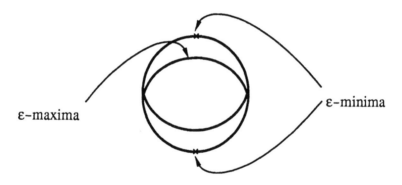

iii) *Similarly, if $\lambda_1 = \lambda_2 > \lambda_3$, then there are two ε–maxima and an equator of ε–minima.*

iv) *If $\lambda_1 > \lambda_2 > \lambda_3$, then there are two ε–minima, two ε–maxima and two ε–saddles.*

(1.25) Some of the equivariant harmonic maps in (1.24) are harmonic morphisms (actually Riemannian submersions up to a constant factor); but others are not. For instance,

Assume $\lambda_1 > \lambda_2 \geq \lambda_3$. Then the equivariant harmonic map $\tilde{\varphi}_{[v]} : SU(2) \to \mathbb{C}P^1$ associated with the ε–minimum $[v] = [1 : 0]$ is a harmonic morphism iff $\lambda_2 = \lambda_3$.

In the terminology of Appendix 1 (38), the case $\lambda_2 = \lambda_3$ is that of the Hopf fibration of a Berger sphere to $\mathbb{C}P^1$.

Proof. By equivariance, it suffices to study the condition of horizontal conformality at the identity of $SU(2)$:

$$d\tilde{\varphi}_{[v]} : su(2) \longrightarrow T_v \,\mathbb{C}P^1$$

$$\xi \longrightarrow d\pi\,((\xi v)_v)\,.$$

Thus ξ is tangent to the fibre iff $(\xi v)_v$ is vertical with respect to $\pi : \mathbb{C}^2 - \{0\} \to \mathbb{C}P^1$, iff

(1.26) $$\|\xi v\|^2 - \frac{|<\xi v, v>|^2}{\|v\|^2} = 0\,.$$

Now write ξ as

$$\xi = \begin{pmatrix} \ell & -\bar{m} \\ m & -\ell \end{pmatrix}$$

where $\ell, m \in \mathbb{C}$, $Re(\ell) = 0$.

Since $[v] = [1:0]$, it is easy to see that (1.26) becomes

$$|\ell|^2 + |m|^2 - |\ell|^2 = 0$$

which implies $m = 0$.

Thus ξ spans the subspace of $su(2)$ tangent to the fibre. From this it is easy to conclude that $\tilde{\varphi}_{[v]}$ is horizontally conformal iff $\lambda_2 = \lambda_3$. //

(1.27) Let $C = diag(\lambda_1, \lambda_2, \lambda_3)$ as above. Then the Ricci quadratic form R_{ij} associated with the left–invariant metric (1.20) is diagonal with respect to the orthonormal basis $\left(e_i = \frac{1}{\sqrt{\lambda_i}}\,\xi_i\right)$; following [Mi], we obtain the principal Ricci curvatures

$$r(e_1) = \{\lambda_2^2\lambda_3^2 - \lambda_1^2[\lambda_2 - \lambda_3]^2\}/\lambda_1\lambda_2\lambda_3$$
$$r(e_2) = \{\lambda_1^2\lambda_3^2 - \lambda_2^2[\lambda_1 - \lambda_3]^2\}/\lambda_1\lambda_2\lambda_3$$
$$r(e_3) = \{\lambda_2^2\lambda_1^2 - \lambda_3^2[\lambda_2 - \lambda_1]^2\}/\lambda_1\lambda_2\lambda_3\,.$$

Therefore simple inspection shows that equivariant harmonic maps as in (1.22) and (1.24) i), ii), iii) exist only if R_{ij} has signature $(+, +, +)$. By way of contrast, we have examples of equivariant harmonic maps as in (1.24) iv) for signatures $(+,+,+), (+,0,0)$ and $(+,-,-)$.

(1.28) There is a natural 1-1 correspondence between G-invariant *indefinite* metrics on G/H and $\mathrm{ad}(H)$-invariant non-degenerate symmetric bilinear forms on $\mathcal{G}/\mathcal{H} = T_{x_0}(M)$. The constructions in this section apply to indefinite metrics without significant modifications. By way of example, we state the

Theorem. *Let G/H be a compact homogeneous space endowed with any G-invariant indefinite metric. Then G/H can be harmonically immersed into some S^n.*

2.   ROTATION HYPERSURFACES IN SPHERES

(2.1) Take $n \geq 3$ and let $SO(n-1)$ act isometrically on S^n by its standard action on $\mathbb{R}^{n-1}$, being constant on the (x_n, x_{n+1})-subspace of $\mathbb{R}^{n+1}$. There are just two orbit types: the principal orbits, which are Euclidean $(n-2)$-spheres; and the singular orbits, which are points. We parametrize the orbit space as follows: let $C \subset S^n$ be the great circle in (x_n, x_{n+1})-space of the fixed points of the $SO(n-1)$-action. Let S_+^2 denote the *closed* hemisphere in (x_{n-1}, x_n, x_{n+1})-space whose boundary $\partial S_+^2 = C$.

The interior points parametrize the principal orbits (there being just one point where such meets Int S_+^2). The boundary C parametrizes the singular orbits. We denote by $\sigma : S^n \to S_+^2 = S^n/SO(n-1)$ the projection onto the orbit space.

(2.2) Introduce polar coordinates (η, θ) on S_+^2, so that $0 \leq \theta \leq 2\pi$ is the angle; and $0 \leq \eta \leq \pi/2$ is the radial coordinate from C to the centre of S_+^2. The canonical metric on S^n descends to the metric $h = d\eta^2 + \cos^2\eta \, d\theta^2$ on S_+^2, with respect to which σ is a Riemannian submersion on $(S^n)^* = \sigma^{-1}(\mathrm{Int}\, S_+^2)$.

Moreover, the volume function (IV (3.17)) is

$$V = V(\eta) = \mathrm{Vol}(S^{n-2}(1)) \sin^{n-2}\eta.$$

Let $\gamma : P \to S_+^2$ be a curve parametrized by arc length s. According to IV (4.8), γ

determines a commutative diagram

(2.3)
$$\begin{array}{ccc} \gamma^{-1}(S^n) = M & \xrightarrow{\varphi} & S^n \\ \rho \downarrow & & \downarrow \sigma \\ \mathbb{R} \supset P & \xrightarrow{\gamma} & S^2_+. \end{array}$$

Corollary IV (4.9) tells us that φ is a minimal immersion iff γ is a geodesic in (S^2_+, h_1), where $h_1 = V^2 h$.

Since the metric h_1 is invariant under the reflection $\theta_0 + \theta \to \theta_0 - \theta$, we conclude that the diameters θ = constant are all geodesics; the corresponding minimal immersions are just totally geodesic hyperspheres in S^n.

By uniqueness, all the geodesics passing through the centre of S^2_+ must be diameters; therefore we can assume that a geodesic γ which is not a diameter has the form $\gamma(s) = (\eta(s), \theta(s))$, with $\eta(s) \neq \pi/2$. Indicating by $\cdot$ differentiation with respect to s, we have

$$\dot{\eta} = \cos\alpha, \quad \dot{\theta} = \frac{\sin\alpha}{\cos\eta};$$

$$\kappa(\gamma) = \dot{\alpha} - \sin\eta\,\dot{\theta}$$

where $\kappa(\gamma)$ is the geodesic curvature of γ; and

$$\frac{d}{d\nu} = -\sin\alpha\,\frac{\partial}{\partial\eta} + \frac{\cos\alpha}{\cos\eta}\,\frac{\partial}{\partial\theta}.$$

According to Corollary IV (4.11), φ is a minimal immersion iff

$$\kappa(\gamma) - \frac{d}{d\nu}\log V = 0;$$

i.e.

(2.4)
$$\dot{\alpha} - \sin\eta\,\dot{\theta} + (n-2)\sin\alpha\,\frac{\cos\eta}{\sin\eta} = 0.$$

We look for *closed* geodesics contained in Int S^2_+; it is clear by uniqueness that no such geodesic can be tangent to a diameter. Therefore, in complete generality we can treat η as a function of θ, where now θ belongs to $\mathbb{R}$. We have

$$\eta' = \frac{d\eta}{d\theta} = \frac{d\eta}{ds}\frac{1}{\dot{\theta}} = \frac{\cos\alpha\cos\eta}{\sin\alpha}$$

and so

$$\frac{V\cos^2\eta}{\sqrt{\cos^2\eta + \eta'^2}} = V\cos\eta\sin\alpha.$$

Consequently,

$$\frac{d}{d\theta}\left[\frac{V\cos^2\eta}{\sqrt{\cos^2\eta+\eta'^2}}\right] = \frac{1}{\dot\theta}\frac{d}{ds}[V\cos\eta\sin\alpha]$$

$$= \frac{V\cos^2\eta\cos\alpha}{\sin\alpha}\left[\dot\alpha - \sin\eta\,\dot\theta + (n-2)\sin\alpha\frac{\cos\eta}{\sin\eta}\right] = 0$$

by (2.4).

That implies the existence of a constant $K > 0$ such that

(2.5) $$\left(\frac{d\eta}{d\theta}\right)^2 = \frac{V^2\cos^4\eta}{K^2} - \cos^2\eta.$$

(2.6) Let $\eta = \eta(\theta)$ be the unique geodesic determined by initial conditions

$$\eta(0) = a, \qquad \eta'(0) = 0$$

for some $a \in (0, \pi/2)$.

Using (2.5), we have immediately

(2.7) $$\left(\frac{d\eta}{d\theta}\right)^2 = \cos^2\eta\left[\frac{\sin^{2(n-2)}\eta\cos^2\eta}{\sin^{2(n-2)}a\cos^2 a} - 1\right]$$

from which we obtain

$$\left(\frac{d^2\eta}{d\theta^2}\right)(0) = (\cot a)[(n-1)\cos^2 a - 1].$$

In particular, we observe that

$$\left(\frac{d^2\eta}{d\theta^2}\right)(0) \leq 0 \quad\text{iff}\quad a \geq \arccos\left(\frac{1}{\sqrt{n-1}}\right) = a_0$$

Case $a = a_0$. A simple inspection of (2.4) shows that $\eta(\theta) \equiv a_0$ is the unique solution such that $\eta(0) = a_0$, $\eta'(0) = 0$. This geodesic is the extremal orbit of the rotation group $SO(2)$ acting on S^2_+; therefore the corresponding minimal immersion is $SO(n-1) \times SO(2)$-invariant. It is just the Clifford torus

(2.8) $$S^{n-2}\left(\sqrt{\frac{n-2}{n-1}}\right) \times S^1\left(\sqrt{\frac{1}{n-1}}\right)$$

embedded in S^n.

Case $a > a_0$. Now $\dfrac{d^2\eta}{d\theta^2}(0) < 0$, so $\theta = 0$ is a local maximum; thus as θ increases from 0, $\eta(\theta)$ decreases: more precisely, inspection of (2.7) tells us that $\eta(\theta)$ will decrease strictly until $\theta = \Omega_a$, where $\eta(\Omega_a) = a'$ satisfies $\sin^{2(n-2)} a' \cos^2 a' = \sin^{2(n-2)} a \cos^2 a$.

To prove the existence of Ω_a, suppose that $\eta(\theta) > a'$ for all $\theta > 0$. This obviously implies that $\lim_{\theta \to +\infty} \eta'(\theta) = 0$ and so, from (2.7), $\lim_{\theta \to +\infty} \eta(\theta) = a'$. Going back to the arc length parametrization, we obtain

$$\lim_{s \to +\infty} \eta(s) = a', \quad \lim_{s \to +\infty} \sin \alpha(s) = 1, \quad \lim_{s \to +\infty} \dot\theta(s) = \frac{1}{\cos a'}.$$

But $a' \ne a_0$, so these contradict (2.4), thus proving the existence of Ω_a.

By (2.7) and the definitions of a' and Ω_a, $\eta'(\Omega_a) = 0$; therefore we can continue $\eta(\theta)$ as a geodesic simply by reflection with respect to the diameter $\theta = \Omega_a$; iterating this procedure at the points $\{i\Omega_a\}_{i \in \mathbb{Z}}$ we conclude that $\eta(\theta)$ defines a closed geodesic iff the period $2\Omega_a$ is a rational multiple of π, where

$$\Omega_a = \int_{a'}^{a} \frac{d\eta}{f(\eta)}$$

and $f^2(\eta)$ denotes the right member of (2.7). Clearly there is a countable infinity of these. On the other hand, any closed geodesic γ must have a local maximum, which we may assume to be taken at $\theta = 0$; thus $\eta(\theta)$ satisfies initial conditions (2.6). That determines (up to rotation in θ) all the closed geodesics of (S_+^2, h_1).

Let

$$A_n = \{a \in (a_0, \pi/2) : (\Omega_a/\pi) \text{ is rational}\} \quad \text{and}$$
$$\varphi_a : M_a \to S^n$$

denote the corresponding minimal immersion.

Then we have the following enumeration

(2.9) Theorem. *The closed minimally immersed hypersurfaces in S^n of cohomogeneity 1 with respect to the standard $SO(n-1)$-action are:*

(i) *The great hyperspheres (totally geodesic), which lie over the diameters of S_+^2.*

(ii) *The Clifford torus (2.8), lying over the $SO(2)$-invariant geodesic $\eta \equiv a_0$.*

(iii) *The hypersurfaces $\varphi_a : M_a \to S^n$ with $a \in A_n$.*

(2.10) Remark. For every $a \in A_n$, M_a is homeomorphic to $S^{n-2} \times S^1$. Clearly the map φ_a is an embedding iff π/Ω_a is integral.

(2.11) Remark. Consider the ellipsoid

$$Q^n(a,b) = \left\{ (x,y) \in \mathbb{R}^{n-1} \times \mathbb{R}^2 : \frac{|x|^2}{a^2} + \frac{|y|^2}{b^2} = 1 \right\}.$$

The $SO(n-1)$-action on $\mathbb{R}^{n+1}$ defined in (2.1) is isometric on $Q^n(a,b)$. A straightforward modification of the analysis of paragraphs (2.1–10) shows that for any choice of a, b, the closed, minimally immersed $SO(n-1)$-invariant hypersurfaces of cohomogeneity 1 in $Q^n(a,b)$ can be enumerated precisely as in Theorem (2.9). An analogous phenomenon is described in detail in the examples following this Remark.

By way of contrast, we will show in Chapters IX, X that the existence of equivariant harmonic maps can be altered drastically by ellipsoidal deformations.

(2.12) Minimal surfaces in $Q^3(a,b)$. Consider the ellipsoid

$$Q^3(a,b) = \left\{ (x,y) \in \mathbb{C} \times \mathbb{C} : \frac{|x|^2}{a^2} + \frac{|y|^2}{b^2} = 1 \right\}$$

parametrized by join coordinates (u,v,r), with $u,v \in S^1$ and $0 \leq r \leq \pi/2$ (as in X, Section 3); its Riemannian metric is

(2.13) $$g = a^2 \sin^2 r\, du^2 + b^2 \cos^2 r\, dv^2 + K^2(r)\, dr^2,$$

where $K^2(r) = a^2 \cos^2 r + b^2 \sin^2 r$.

For relatively prime integers $m \geq k \geq 1$ we define the isometric action $G_{m,k}$ of S^1 on $Q^3(a,b)$ by

(2.14) $$\gamma \cdot (x,y) = (e^{im\gamma} x, e^{ik\gamma} y) \quad \text{for } \gamma \in S^1.$$

In terms of (u,v,r) this action is expressed by

(2.15) $$\gamma \cdot (u,v,r) = (u + m\gamma, v + k\gamma, r).$$

We denote the orbit space $Q = Q^3(a,b)/G_{m,k}$, with projection map $\sigma : Q^3(a,b) \to Q$.

Let $I^2 = \{(\theta, \eta) : 0 \leq \theta < 2\pi, 0 \leq \eta \leq \pi\}$; and define the map $\psi : I^2 \to Q^3(a, b)$ by $\psi(\theta, \eta) = (0, \theta, \eta/2)$. We observe that ψ maps each point of $\mathrm{Int}(I^2)$ to exactly one $G_{m,k}$–orbit, so $\sigma \circ \psi : I^2 \to Q$ serves a coordinate chart; furthermore,

(2.16) $$\psi_* \frac{\partial}{\partial \theta} = \frac{\partial}{\partial v}, \quad \psi_* \frac{\partial}{\partial \eta} = \frac{1}{2} \frac{\partial}{\partial r}.$$

The orbit through the point $\psi(\theta, \eta)$ is given by the curve $x(t) = (mt, \theta + kt, \eta/2)$; $0 \leq t < 2\pi$, starting at $x(0) = \psi(\theta, \eta)$. The tangent vector $\dot{x}(0) = m \frac{\partial}{\partial u} + k \frac{\partial}{\partial v}$ has length ^{2} given (via (2.13)) by

(2.17) $$|\dot{x}(0)|^2 = m^2 a^2 \sin^2(\eta/2) + k^2 b^2 \cos^2(\eta/2).$$

The horizontal component of a vector $w \in T_{x(0)}(Q^3(a, b))$ is

(2.18) $$w^H = w - \frac{g(w, \dot{x}(0))\dot{x}(0)}{|\dot{x}(0)|^2}.$$

Using (2.13) and (2.16–2.18),

$$\left(\psi_* \frac{\partial}{\partial \theta}\right)^H = \frac{1}{|\dot{x}(0)|^2} \cdot \left\{-mkb^2 \cos^2(\eta/2) \frac{\partial}{\partial u} + m^2 a^2 \sin^2(\eta/2) \frac{\partial}{\partial v}\right\}$$

$$\left(\psi_* \frac{\partial}{\partial \eta}\right)^H = \frac{1}{2} \frac{\partial}{\partial r}.$$

Thus we are able to express in the coordinates (θ, η) the metric h with respect to which $\sigma : (Q^3(a, b), g) \to (Q, h)$ is a Riemannian submersion:

(2.19)
$$h = \frac{a^2 b^2 m^2 \sin^2 \eta}{[m^2 a^2 \sin^2(\eta/2) + b^2 k^2 \cos^2(\eta/2)]} d\theta^2 + [b^2 \sin^2(\eta/2) + a^2 \cos^2(n/2)] d\eta^2.$$

Thus (Q, h) is an *ovaloid*. (Note that $Q = S^2$ (up to homothety) iff $a = b$ and $m = k$.)

Moreover, up to a constant the volume function is

(2.20) $$V = V(\eta) = m^2 a^2 \sin^2(\eta/2) + b^2 k^2 \cos^2(\eta/2).$$

Again, according to Corollary IV (4.9), there is a 1-1 correspondence between closed geodesics in (Q, h_1), where

$$h_1 = V^2 h = a^2 b^2 m^2 \sin^2 \eta \, d\theta^2 + D^2(\eta) \, d\eta^2$$

with

$$D^2(\eta) = m^2a^2b^2 \sin^4(\eta/2) + k^2a^2b^2 \cos^4(\eta/2) + (m^2a^4 + k^2b^4)\sin^2(\eta/2)\cos^2(\eta/2),$$

and closed, cohomogeneity one $G_{m,k}$-invariant minimal surfaces in $Q^3(a,b)$.

The analysis now follows the lines of the first part of this Section: We observe that the metric h_1 is invariant by rotations in the variable θ. Therefore

(2.21) The longitudes (θ = const.) are all closed geodesics. They lie under closed minimal surfaces $T_{m,k}$.

No other geodesic γ is tangent to a longitude (by uniqueness); in particular, γ does not pass through either pole ($\eta = 0, \pi$) of the ovaloid. Consequently, γ can be expressed as a function $\eta = \eta(\theta)$, provided that θ varies over all $\mathbb{R}$. We observe that γ is closed iff $\eta(\theta)$ is a periodic function whose period is a rational multiple of π.

With initial conditions

$$\eta(0) = c, \qquad \eta'(0) = 0,$$

a computation as above (compare with (2.7)) leads us to

(2.22) $$\left(\frac{d\eta}{d\theta}\right)^2 = \frac{m^2a^2b^2 \cos^2 \eta}{D^2(\eta)} \left[\frac{\cos^2 \eta}{\cos^2 c} - 1\right]$$

and

(2.23) $$\frac{d^2\eta}{d\theta^2}(0) = \frac{m^2a^2b^2}{2} \frac{\sin 2c}{D^2(c)}.$$

If $c = \pi/2$, then $\eta \equiv \pi/2$. The corresponding minimal surface is the Clifford torus (if $a = b$).

If $c > \pi/2$, the analysis follows the same pattern as in the case $a > a_0$ above. In particular, $\eta(\theta)$ has critical values c and $\pi - c$. It oscillates between the latitudes $\eta = c$ and $\eta = \pi - c$ with period $2\Omega_c$, where

$$\Omega_c = \int_c^{\pi-c} \frac{d\eta}{f(\eta)}$$

and $f^2(\eta)$ denotes the right member of (2.22).

Let

$$A_{m,k} = \{c \in (\pi/2, \pi) : (\Omega_c/\pi) \text{ is rational}\} \cup \{\pi/2\}.$$

Since Ω_c is a continuous, non-constant function of c, we conclude that the set $A_{m,k}$ has a countable infinity of elements.

(2.24) We denote by $T_{m,k,c} = \sigma^{-1}(\gamma_c)$ the minimal surface in $Q^3(a,b)$ corresponding to $c \in A_{m,k}$. We have proved

(2.25) **Theorem.** *For relatively prime integers $m \geq k \geq 1$ the $G_{m,k}$-invariant, cohomogeneity 1 minimal surfaces in $Q^3(a,b)$ are (up to congruences):*

i) The $T_{m,k}$ of (2.21), which lie over the longitudes of Q.

ii) The $T_{m,k,c}$ of (2.24): for fixed $m > k$ the family $T_{m,k,c}$ is countably infinite; and T_{m,k,c_1} is congruent to $T_{m,k,c}$ iff $c_1 = c$.

(2.26) **Remark.** For any choice of a, b (in particular for $a = b = 1$), the actions of $G_{m,k}$ of type (2.14), together with the action (2.1) with $n = 3$, are (up to conjugacy) the only isometric actions on $Q^3(a,b)$ with principal orbits of codimension 2.

3. CONSTANT MEAN CURVATURE ROTATION HYPERSURFACES IN $\mathbb{R}^n$

(3.1) Let $SO(n-1)$ operate on $\mathbb{R}^n$ by its standard action on the first $(n-1)$ coordinates, and trivially on the last.

It is clear that the orbit space Q can be parametrized by $Q = \{(x,y) \in \mathbb{R}^2 : y \geq 0\}$; moreover, the projection $\sigma : \mathbb{R}^n \to Q$ is a Riemannian submersion provided that Q is given the flat metric.

The orbits associated with the interior points are Euclidean $(n-2)$-spheres; the singular orbits $(x, 0)$ are points.

Up to a constant, the volume function is

$$V = V(y) = y^{n-2}.$$

(3.2) Let $P \subset \mathbb{R}$ be an interval and $\gamma : P \to Q$ a curve parametrized by arc length: $\gamma(s) = (x(s), y(s))$. As in IV (4.8) the pull-back construction on γ produces a commutative diagram

$$\gamma^{-1}(\mathbb{R}^n) = M \xrightarrow{\varphi} \mathbb{R}^n$$
$$\rho \downarrow \qquad \downarrow \sigma$$
$$\mathbb{R} \supset P \xrightarrow{\gamma} Q.$$

According to IV (4.11), $\varphi : M \to \mathbb{R}^n$ is an immersion of constant mean curvature H iff γ satisfies

(3.3) $$(n-1)H = \kappa(\gamma) - \frac{d}{d\nu} \log V.$$

Since Q is flat, the geodesic curvature is

$$\kappa(\gamma) = \dot{x}\,\ddot{y} - \dot{y}\,\ddot{x} ;$$

and

$$\frac{d}{d\nu} \log V = (n-2)\frac{\dot{x}}{y}.$$

Therefore (3.3) becomes

(3.4) $$\dot{x}\,\ddot{y} - \dot{y}\,\ddot{x} = (n-2)\frac{\dot{x}}{y} + (n-1)H.$$

Because γ is parametrized by arc length, $\dot{x}^2(s) + \dot{y}^2(s) = 1$, so $\dot{x}\,\ddot{x} + \dot{y}\,\ddot{y} = 0$, which can be rearranged in the form

(3.5) $$\frac{\ddot{x}}{\dot{y}} - \ddot{x}\,\dot{y} + \dot{x}\,\ddot{y} = 0.$$

(3.6) Let $J = y^{n-2}\dot{x} + H\,y^{n-1}$. Then

(3.7) $$\frac{dJ}{ds} = \dot{y}\,y^{n-2}\left[\frac{\ddot{x}}{\dot{y}} + (n-2)\frac{\dot{x}}{y} + (n-1)H\right].$$

If $\gamma(s) = (x(s), y(s))$ is a solution of (3.4), then J is constant along γ. In fact, if we substitute (3.4) in (3.7) we get

$$\frac{dJ}{ds} = \dot{y}\,y^{n-2}\left[\frac{\ddot{x}}{\dot{y}} - \ddot{x}\,\dot{y} + \dot{x}\,\ddot{y}\right] = 0$$

by (3.5).

In other words, J is a *prime integral* of equation (3.4). That fact can be used to obtain a very good qualitative description of the solutions of (3.4), as follows:

(3.8) Let $f_H(y) = f(y) = H y^{n-1} - y^{n-2}$, $y \geq 0$, $H > 0$. Then $f(0) = f\left(\frac{1}{H}\right) = 0$; $f(y)$ is negative on $\left(0, \frac{1}{H}\right)$ and attains its minimum at $y^* = (n-2)/[(n-1)H]$. Moreover, given $0 > c > f(y^*)$, there exist exactly two points $m_c < M_c$ such that $f(m_c) = f(M_c) = c$.

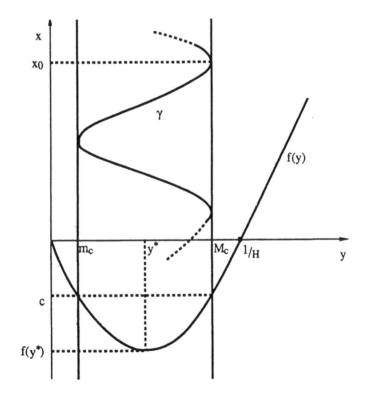

Figure 3.8

We prove

(3.9) **Proposition.** *Let $H > 0$ be fixed. If $0 > c > f(y^*)$, there exists a unique* (up to translation in the x-direction) solution curve γ of (3.4) such that $J \equiv c$ along γ. Moreover, γ is a wave-like periodic curve which oscillates between the lines $y = m_c$ and $y = M_c$.

Proof. Since $\dot{x} \geq -1$ and $y \geq 0$

$$J = H\, y^{n-1} + \dot{x}\, y^{n-2} \geq f(y)\,.$$

Therefore $J = c$ clearly implies

(3.10) $$m_c \leq y \leq M_c\,.$$

Now fix $x_0 \in \mathbb{R}$ and consider the unique solution γ of (3.4) determined by initial conditions

(3.11) $$x(0) = x_0,\quad y(0) = M_c,\quad \dot{x}(0) = -1,\quad \dot{y}(0) = 0\,.$$

We have

$$J\big|_{s=0} = f(y(0)) = c\,.$$

Thus $J \equiv c$ along γ. We observe that $y \equiv M_c$ is *not* a solution of (3.4). Therefore (3.10) tells us that, as s increases from zero, γ must head toward the interior of the strip $[m_c, M_c]$ (see Figure 3.8 above); i.e., $\dot{y}(s) < 0$ for small $s > 0$.

We claim that $\dot{y}(s) < 0$ until s reaches a point s_1 such that

(3.12) $$y(s_1) = m_c,\quad \dot{y}(s_1) = 0 \quad \text{and} \quad \dot{x}(s_1) = -1\,.$$

Proof of the claim. Inspection of the condition $J = c$ and Figure 3.8 shows that $\dot{y} = 0$ can occur only if $y = m_c$ and $\dot{x} = -1$. Thus it only remains to prove the existence of s_1 such that $\dot{y}(s_1) = 0$; suppose that $\dot{y}(s) < 0$ for all $s > 0$. Since $y(s)$ is contained in $[m_c, M_c]$, $\lim_{s \to +\infty} \dot{y}(s) = 0$, which implies $\lim_{s \to +\infty} \dot{x}(s) = \pm 1$. Therefore γ has a vertical asymptotic line. Because $J \equiv c$ along γ, the only possible asymptote is $y = m_c$. Now $y \equiv m_c$ is *not* a solution of (3.4), which leads easily to a contradiction, and establishes (3.12).

Now observe that (3.4) is invariant under translations along the x-direction; thus we can continue γ by reflection along the line $x = x(s_1)$. Iterating this procedure, we conclude that γ is a wave-like periodic curve (of period $2s_1$), as required. Moreover, let now γ be any solution on which $J \equiv c$. The argument above shows that γ must touch either the line $y = m_c$ or the line $y = M_c$ with vertical tangent. Thus we conclude by uniqueness (up to orientation) that γ is translationally equivalent to the solution determined by (3.11). //

By way of summary: The solutions of Proposition (3.9) *provide constant mean curvature embeddings* of $S^{n-2} \times \mathbb{R}$ into $\mathbb{R}^n$; for $n = 3$ these are *unduloids* (as in (3.15) (iv) below). Moreover, such maps factor to give constant mean curvature embeddings $S^{n-2} \times S^1 \to \mathbb{R}^n$; the associated Gauss maps $S^{n-2} \times S^1 \to S^{n-1}$ are harmonic with respect to the induced metrics, by II (1.10).

The other choices of H and c can be treated by similar methods to give

(3.13) Proposition. *Up to equivalence of translations in the x-direction, the global solutions of (3.4) are uniquely characterized by $J = c$.*

(3.14) In case $n = 3$ we have the classical classification of Delaunay: A *roulette of a conic C* is the curve in a plane traced by a focus of C as C rolls along a straight line, called the axis.

(3.15) Theorem. *The complete immersed surfaces of revolution in $\mathbb{R}^3$ with constant mean curvature are precisely those obtained by rotating about their axes the roulettes of a conic.*

Here is the list, starting with two degenerate cases:

(i) The right circular cylinder has curvature 0 and constant mean curvature $\neq 0$.

(ii) The sphere has constant curvature $1/R^2$ and constant mean curvature $1/R$.

(iii) The catenoid is generated by rotating about its axis the roulette of a parabola (that latter being a catenary). It has variable curvature and 0 mean curvature.

(iv) and (v) The unduloid and nodoid, which are generated by rotating about their axes the roulettes of an ellipse and of a hyperbola respectively. They have variable curvature and constant mean curvature $\neq 0$.

We must identify the equation of the roulette with an equation of type $J = c$. That is best done case by case. //

(3.16) Suppose that a solution γ can be expressed in the form $y = y(x)$ (note that this is always the case for the solutions of Proposition (3.9)). In this case $J = c$ takes the form

(3.17) $$\frac{y^{n-2}(x)}{\sqrt{1+y'^2(x)}} + H\, y^{n-1}(x) = c$$

which yields

$$x = \int \left[\left(\frac{y^{n-2}}{c - H\, y^{n-1}} \right)^2 - 1 \right]^{-1/2} dy\,.$$

This integral can be calculated in terms of hyperelliptic functions.

(3.18) Here is an alternative way to obtain the prime integral (3.17): we consider hypersurfaces of revolution in $\mathbb{R}^n$ determined by functions $y(x) = y : [a, b] \to \mathbb{R}$, with fixed volumes of revolution

$$V(y) = \text{Vol } (B^{n-1}(1)) \int_a^b y^{n-1}(x)\, dx \, ;$$

and extremize their lateral area

$$A(y) = \text{Vol } (S^{n-2}(1)) \int_a^b \left\{ y^{n-2} \sqrt{1 + y'^2(x)} \right\} dx \, .$$

For such an isoperimetric problem, Lagrange's method of multipliers leads us to the Euler–Lagrange equation of the functional

$$\begin{aligned} W(y) &= V(y) + \mu\, A(y) \\ &= \text{Vol } (B^{n-1}(1)) \int_a^b \left\{ y^{n-1}(x) + (n-1)\mu\, y^{n-2}(x) \left[\sqrt{1 + y'^2(x)} \right] \right\} dx \\ &= \text{Vol } (B^{n-1}(1)) \int_a^b L(y, y')\, dx \, . \end{aligned}$$

Since L does not depend explicitly on x, the Hamiltonian $H = L - y' \dfrac{\partial L}{\partial y'}$ is constant along solutions; that is (3.17) with $H = 1/(n-1)\,\mu$.

Notes and comments

(4.1) Theorem (1.4) is a special case of a theorem of Smith [Sm1]. It is an example of symmetric criticality, as described by Palais [P].

(4.2) Hsiang [H1] announced a version of Theorem (1.13). The main ideas of a proof – based on the characterization of equivariant minimal immersions as the extremals of the volume function – were given in [HL]; and fully presented in [W]. Takahashi [Ta] gave a proof in case G/H is irreducible (i.e., the isotropy representation of H on the tangent space $T_0(G/H)$ is irreducible). More related to Theorem (1.13) can be found in [dCW], using spherical harmonics (see also Chapter VIII, Section 1).

Apparently it is still unknown whether every compact Riemannian manifold can be minimally immersed in some S^n.

(4.3) Much of Section 1 is taken from Rawnsley [Ra]; in particular, Proposition (1.16) and Application (1.19–22) are due to him.

The matrices ξ_1, ξ_2, ξ_3 of (1.19) together with the identity I are the *Pauli spin matrices*. They arise naturally from the following construction: We identify

$$S^3 \xrightarrow{f} SU(2)$$
$$\cap$$
$$\mathbb{C}^2$$

where $f(z, w) = \begin{pmatrix} z & -\bar{w} \\ w & \bar{z} \end{pmatrix}$. Then f carries the base vector $(1,0,0,0), (0,1,0,0)$, $(0,0,1,0), (0,0,0,1)$ to I, ξ_1, ξ_2, ξ_3 respectively.

The standard inner product on the Lie algebra

$$su(2) = \left\{ \begin{pmatrix} i\theta & -\bar{w} \\ w & -i\theta \end{pmatrix} : \theta \in \mathbb{R}, w \in \mathbb{C} \right\}$$

is $<X, Y> = -\text{Trace } XY$; thus the norm²

$$\left\| \begin{pmatrix} i\theta & -\bar{w} \\ w & -i\theta \end{pmatrix} \right\|^2 = \theta^2 + |w|^2 .$$

That inner product determines a bi-invariant Riemannian metric on the special unitary group $SU(2)$ with respect to which f is an isometry. Its associated connection is given by

$$\nabla_X Y = \frac{1}{2}[X, Y];$$

the Laplacian operating on functions

$$\Delta^{SU(2)} = -\sum_{i=1}^{3} \xi_i \circ \xi_i .$$

Let Y, U, V denote the orthonormal 3-frame field on S^3 determined by $((df)^{-1}(\xi_i))$, $1 \le i \le 3$, then

$$\Delta^{S^3} = Y \circ Y - \nabla_Y Y + U \circ U - \nabla_U U + V \circ V - \nabla_V V .$$

The Hopf fibration can be viewed (compare VIII (2.8)) as the coset map

$$\begin{array}{ccc} SU(2) & \longrightarrow & SU(2)/S(U(1) \times U(1)) \\ \| & & \| \\ S^3(1) & & S^2(\tfrac{1}{2}) \end{array}$$

where $S(U(1) \times U(1)) = \left\{ \begin{pmatrix} e^{i\theta} & 0 \\ 0 & e^{-i\theta} \end{pmatrix} : 0 \leq \theta \leq 2\pi \right\}$. We also observe that any $SU(2)$–invariant metric on $S^3(1)$ coincides with one of the Berger metrics (Appendix 1 (38)) [BB].

(1.17) can be found in [Her]; and (1.18) in [Gu].

(4.4) In following the presentation in [HL] of Otsuki's Theorem (2.9), we clear the way to a case–by–case enumeration of the cohomogeneity one minimal immersions of hypersurfaces in S^n. See also [dCD]. Amongst those, the product action of $S^1 \times S^1 \times S^1$ on S^5 in $\mathbb{R}^2 \times \mathbb{R}^2 \times \mathbb{R}^2 = \mathbb{R}^6$ has important special features. Its orbit space Q is a 2–spherical triangle in the first octant of $\mathbb{R}^3$:

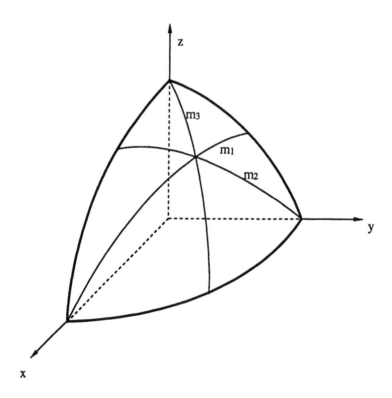

Figure 4.5

Take parameters (η, θ) with $0 \leq \eta \leq \pi/2$, $0 \leq \theta < 2\pi$, so Q is parametrized by

$$x = \cos \eta \cos \theta$$
$$y = \cos \eta \sin \theta$$
$$z = \sin \eta .$$

Then the metric on Q is $h = d\eta^2 + \cos^2 \eta \, d\theta^2$, and the volume function $V = \cos^2 \eta \sin \eta \cos \theta \sin \theta$.

Interior points of Q correspond to submanifolds of S^5 of type $S^1 \times S^1 \times S^1$; the three boundary open edges to those of type $S^1 \times S^1$ (deleting one factor); the three vertices to those of type S^1 (deleting two factors).

There are countably many closed solution curves in Int Q, obtained by successive reflections in the three meridians m_1, m_2, m_3 of Q. An embedded closed solution C is obtained, with equation $\eta = \eta(\theta)$, starting at $a = \eta(\pi/4)$ with $\eta'(\pi/4) = 0$; and orthogonal to $\theta = \pi/4$.

Thus C lies *under a torus* T^4 *minimally embedded in* S^5.

Apparently it is unknown whether there are tori T^{n-1} minimally embedded in S^n for $n \neq 3, 5$. See also [HH].

(4.6) Theorem (2.25) was proved in [HL] with $a = b = 1$. Lawson [L2] showed that

$$T_{m,k} = \{(z, w) \in S^3 \subset \mathbb{C}^2 : Re(z^k \bar{w}^m) = 0\} .$$

It is non-orientable iff $2|mk$; and has Euler characteristic 0, (i.e., topologically $T_{m,k}$ is either a torus or a Klein bottle). $T_{m,k}$ is a ruled surface, algebraic of degree $m + k$.

(4.7) Here is an outline of Lawson's beautiful construction [L2] of closed minimal surfaces in the Euclidean 3–sphere S^3. The limitations imposed on this monograph preclude presentations of his proof – which is based on the existence and qualitative properties of the solution to Plateau's problem.

Choose points P_1, P_2 and Q_1, Q_2 on two orthogonal geodesic circles of S^3. With integers $m \geq k \geq 1$, suppose

$$dist \, (P_1, P_2) = \pi/(k + 1), \quad dist \, (Q_1, Q_2) = \pi/(m + 1) .$$

Let Γ be the closed polygon $P_1Q_1P_2Q_2$ indicated in Figure 4.7:

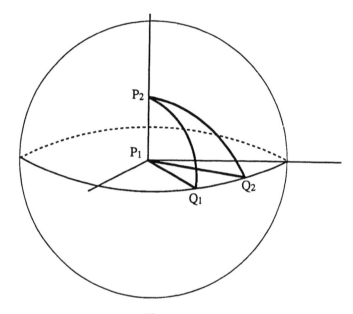

Figure 4.7

The solution to Plateau's problem provides a map $\psi : D \to S^3$ of the closed unit disc $D \subset \mathbf{R}^2$, carrying $\partial D \to \Gamma$. ψ is a minimal conformal embedding, analytic except possibly at the points of ∂D corresponding to the vertices of Γ. Iterated reflections of the image $\psi(D)$ across the geodesic arcs of Γ produce a *compact orientable minimal surface* $M_{m,k}$ *embedded in* S^3. An application of the Gauss–Bonnet theorem shows that its Euler charcateristic $\chi(M_{m,k}) = 2(1 - mk)$.

If we had started with the polygon $\Gamma = P_2Q_2P_1(-Q_2)Q_1$ then, provided that k is odd, the construction can be modified to produce *a compact non–orientable surface* $M_{m,k}$ *minimally immersed in* S^3 – with $\chi(M_{m,k}) = 1 - mk$. Thus we see how to represent every closed surface as a minimally immersed submanifold of S^3 – with the single exception of the real projective plane P, which is prohibited by III (2.14).

Composing $M \xrightarrow{\psi} S^3 \xrightarrow{i} \mathbf{R}^4$ gives an immersion $\varphi = i \circ \psi : M \to \mathbf{R}^4$ with constant mean curvature, by II (1.3). There are related constructions in [KPS].

A different reflection argument has been used by Lemaire [Le] to construct harmonic maps $\varphi : M \to S$ of low degree.

(4.8) For $n = 3$ the classification Theorem (3.15) was established by Delaunay [De] – with an appendix by Sturm, giving (3.18). For an exposition, together with applications to harmonic maps, see [Ee]. The rolling construction was generalized to the case $n \geq 4$ by Hsiang–Yu [HY], based on their version of Proposition (3.9). Hano and Nomizu have produced a version for a surface of revolution in Minkowski space [HN].

CHAPTER VI. MINIMAL EMBEDDINGS OF HYPERSPHERES IN S^4

INTRODUCTION

The theorem of Almgren–Calabi (III (2.9)) shows that any minimal embedding of the 2–sphere in S^3 is an equator. In this Chapter we construct a countable infinity of noncongruent minimal embeddings of the 3–sphere in S^4. These were discovered by W–Y. Hsiang [H2]; we present his proof. As in Chapters IV and V, the context is that of equivariant differential geometry. We consider $SO(2) \times SO(2)$–equivariant embeddings of cohomogeneity 1; the minimality condition reduces to an O.D.E. for a generating curve γ in the orbit space $Q = S^4/SO(2) \times SO(2)$. The minimal embeddings are in bijective correspondence with a family of solution curves γ which start and end at the boundary of Q.

Because the O.D.E. is singular on ∂Q, the existence of solutions starting at the boundary requires special proof.

A similar difficulty arises in the study of c.m.c. immersions of S^{n-1} in $\mathbb{R}^n$, which are the object of Chapter VII; in Section 2 we provide a unified treatment of the required existence.

As emphasized in I (2.3), the domain of a minimal immersion $\varphi : M \to (N, h)$ always has the induced metric φ^*h. Thus the 3-spheres obtained from Hsiang's theorem are not Euclidean spheres, unless they are totally geodesic.

1. DERIVATION OF THE EQUATION AND MAIN THEOREM

(1.1) Proposition. *Let $SO(2) \times SO(2)$ act orthogonally on S^4 via the standard representation ((1.2) below). Then the orbit space $Q = S^4/SO(2) \times SO(2)$ is the spherical sector*

$$Q = \{(\eta, \theta): \ 0 \leq \eta \leq \pi, \ 0 \leq \theta \leq \pi/2\} \text{ in } S^2, \text{ with metric}$$
$$h = \sin^2 \eta \, d\theta^2 + d\eta^2 \, .$$

Moreover, (up to a constant) the volume function is

$$V = V(\eta, \theta) = \sin^2 \eta \sin 2\theta \, .$$

Proof. Parametrize points $z \in S^4$ as follows:

$$z = [\sin \eta \cos \theta \, e^{i\varphi}, \sin \eta \sin \theta \, e^{i\xi}, \cos \eta]$$

where $0 \leq \eta \leq \pi$, $0 \leq \theta \leq \pi/2$, $0 \leq \varphi, \xi < 2\pi$.

An element $(e^{i\beta}, e^{i\gamma}) \in SO(2) \times SO(2)$ acts on S^4 by

(1.2) $\quad (e^{i\beta}, e^{i\gamma}) \cdot z = \left[\sin \eta \cos \theta \, e^{i(\varphi+\beta)}, \sin \eta \sin \theta \, e^{i(\xi+\gamma)}, \cos \eta\right]$.

Letting β, γ range over $[0, 2\pi)$, we deduce immediately that the orbit space

$$Q = [\sin \eta \cos \theta, \sin \eta \sin \theta, \cos \eta] \subset \mathbf{C} \times \mathbf{C} \times \mathbf{R}$$

where $0 \leq \eta \leq \pi$, $0 \leq \theta \leq \pi/2$.

Moreover, inspection of (1.2) tells us that Q *is orthogonal to the orbits* of the $SO(2) \times SO(2)$–action; therefore the metric h which makes the projection $\sigma : S^4 / SO(2) \times SO(2) \to Q$ a Riemannian submersion is precisely the metric of Q as a subset of S^4 – that is $h = \sin^2 \eta \, d\theta^2 + d\eta^2$, as required. The explicit expression for the volume function also follows by inspection of (1.2). //

The orbit space Q may be visualized in the following figure:

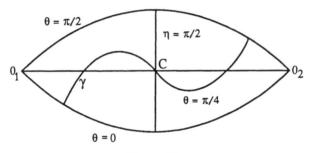

Figure 1.3

If w is an *interior* point, then the orbit $\sigma^{-1}(w) \subset S^4$ is homeomorphic to $S^1 \times S^1$.

If $w \neq O_1, O_2$ is a *boundary* point, then $\sigma^{-1}(w)$ is homeomorphic to S^1; the orbits associated with O_1 and O_2 are just points. Therefore

(1.4) Proposition. *Let γ be a curve in Q which starts at a point $(\eta_0, 0)$ and ends at $(\eta_1, \pi/2)$ without self–intersections, $0 < \eta_0, \eta_1 < \pi$. Then $\sigma^{-1}(\gamma) \subset S^4$ is an embedded 3–sphere.*

For instance, the pre–image of the η–bisector is the unique $SO(2) \times SO(2)$–invariant equator in S^4.

The θ–bisector corresponds to the suspension of the Clifford torus; its pre–image in S^4 is singular at the two poles which lie over O_1, O_2.

It is worth noticing that the geometry of the orbit space Q is symmetric with respect to both the η–bisector and the θ–bisector; therefore it is also symmetric with respect to their intersection point C.

(1.5) Let $\gamma(s) = (\eta(s), \theta(s))$ be a curve in Q parametrized by arc length s. According to formula IV (4.11) (with $H = 0$), $\sigma^{-1}(\gamma)$ is minimal in S^4 iff

(1.6) $$k(\gamma) - \frac{d}{d\nu} \log V = 0.$$

(1.7) **Main Theorem** (W–Y. Hsiang). *For each positive odd integer $(2i+1)$, there exists an $SO(2) \times SO(2)$–invariant minimal embedding of a 3–sphere in S^4 whose generating curve γ in Q is centrally symmetric with respect to the centre C and intersects the θ–bisector at exactly $(2i + 1)$ points.*

According to (1.4) and (1.5), the proof amounts to showing that the O.D.E. (1.6) admits suitable solution curves γ which start at $\theta = 0$ and end at $\theta = \pi/2$ without self–intersections; the required analysis is carried out in Sections 2, 3 and 4 below. Here we derive the explicit form of (1.6):

(1.8) By definition of arc length s (with respect to the metric h)

$$\dot\eta^2 + \dot\theta^2 \sin^2 \eta = 1$$

where $\cdot$ denotes d/ds.

We introduce the angle α between the curve γ and the radial direction $\frac{\partial}{\partial \eta}$ by

(1.9) $$\dot\eta = \cos \alpha, \qquad \dot\theta = \sin \alpha / \sin \eta.$$

(1.10) **Lemma.** *The minimality equation (1.6) is equivalent to*

(1.11) $$\begin{cases} \dot\alpha = -3 \cot \eta \sin \alpha + 2 \cot(2\theta) \cos \alpha / \sin \eta \\ \dot\eta = \cos \alpha \\ \dot\theta = \sin \alpha / \sin \eta. \end{cases}$$

Proof. We prove that

(1.12) $$k(\gamma) = \dot\alpha + \cos \eta \, \dot\theta$$

(1.13) $$\frac{d}{d\nu} \log V = -2 \left\{ \cos \eta \, \dot\theta + \frac{\cot(2\theta) \dot\eta}{\sin \eta} \right\}.$$

Then (1.6), (1.9), (1.12) and (1.13) together lead immediately to (1.11).

Proof of (1.12). By definition of the geodesic curvature $k(\gamma)$,

(1.14) $\qquad k(\gamma) = <\nu, \ \tau(\gamma)>_h = <\nu, \nabla_{\partial/\partial_s} t>_h \qquad$ where

(1.15) $$t = \dot{\eta}\frac{\partial}{\partial\eta} + \dot{\theta}\frac{\partial}{\partial\theta}$$

is the unit tangent to γ, and

(1.16) $$\nu = -\sin\eta \ \dot{\theta} \frac{\partial}{\partial\eta} + \{\dot{\eta}/\sin\eta\}\frac{\partial}{\partial\theta}$$

is the unit normal.

A straightforward computation with covariant derivatives shows

(1.17) $$\nabla_{\partial/\partial_s} t = [\ddot{\eta} - \dot{\theta}^2 \sin\eta \cos\eta]\frac{\partial}{\partial\eta}$$
$$+ [\ddot{\theta} + 2(\cot\eta)\dot{\eta}\,\dot{\theta}]\frac{\partial}{\partial\theta} \ .$$

Now put (1.17) and (1.16) into (1.14), and use (1.8) and (1.9) to produce (1.12).

Proof of (1.13). This follows by differentiating V in (1.1) in the direction ν, using (1.16). //

(1.18) **Remark.** We can rewrite (1.11) in the form

(1.11)' $\qquad \ddot{\eta} - 3\cos\eta\sin\eta \ \dot{\theta}^2 + 2\cot(2\theta)\dot{\theta}\,\dot{\eta} = 0\ ,$

using (1.9).

From now on, when we say that γ is a solution curve we mean that γ is either a solution of (1.11) or of (1.11)', according to the context.

2. EXISTENCE OF SOLUTIONS STARTING AT THE BOUNDARY

(2.1) The O.D.E. (1.11) is clearly singular at the boundary of Q; since the singularities $\theta = 0$ and $\theta = \pi/2$ are of the same type, we limit ourselves to stating results for solutions starting at $\theta = 0$. Moreover, in order to include the case of c.m.c. immersions of Chapter VII below, we treat the more general case

(2.2) $\qquad \begin{cases} \dot{\alpha} = H(p+q-1) - (p+q-1)\dfrac{G'(\eta)}{G(\eta)}\sin\alpha - \dfrac{K(\theta)}{G(\eta)}\cos\alpha \\ \dot{\eta} = \cos\alpha \\ \dot{\theta} = \sin\alpha/G(\eta) \end{cases}$

where $K(\theta) = (p-1)\tan\theta - (q-1)\cot\theta$ and $G: [0, b) \to \mathbf{R}, b \in (0, +\infty) \cup \{+\infty\}$, is an *analytic* function which is positive on $(0, b)$; $H \in \mathbf{R}, p, q \geq 2$.

Note that (2.2) is equivalent to

(2.2)' $\quad \ddot{\eta} + H(p+q-1)G(\eta)\dot\theta - (p+q-1)G'(\eta)G(\eta)\dot\theta^2 - K(\theta)\dot\theta\,\dot\eta = 0$.

(2.3) **Remarks.** i) (2.2) coincides with (1.11) if $p = q = 2$, $G(\eta) = \sin\eta$, $H = 0$.

ii) The O.D.E. (2.2) arises in the study of $SO(p) \times SO(q)$–invariant immersions of cohomogeneity 1 with c.m.c. H in the rotationally symmetric manifold

$$M^{p+q} = \left(S^{p+q-1} \times [0, b), G^2(\eta)\, g_{S^{p+q-1}} + d\,\eta^2\right).$$

We note that α measures the angle between the generating curve $\gamma = (\eta, \theta)$ and the radial direction $\partial/\partial\eta$. In particular, (2.2) admits the trivial solutions $\alpha \equiv \pi/2, \eta \equiv \bar\eta$; these are distance spheres centred at the pole of M^{p+q}, with c.m.c. $H = G'(\bar\eta)/G(\bar\eta)$.

Here is the main result of this Section:

(2.4) **Proposition.** *For every $0 < a < b$ there exist $\varepsilon > 0$ and a unique solution $\gamma_a(s)$ of (2.2) for $0 \leq s < \varepsilon$, with initial data $\eta(0) = a$, $\theta(0) = 0$. Moreover, $\gamma_a(s)$ is orthogonal to the boundary $\theta = 0$ (i.e., $\dot\eta(0) = 0$) and depends analytically on the parameter a as long as $\delta \leq a \leq b - \delta$, for any $\delta > 0$.*

First we need a preliminary

(2.5) **Lemma.** *If $\gamma_a(s)$ is a solution with initial data $\eta(0) = a, \theta(0) = 0$, then $\dot\eta(0) = 0$.*

Proof of the Lemma. From the third equation in (2.2) we have

$$\dot\theta(0) = \sin\alpha(0)/G(a)$$

which implies

$$\theta(s) \simeq \left(\frac{\sin\alpha(0)}{G(a)}\right) s \quad \text{for} \quad s \simeq 0.$$

We substitute this approximation, together with $K(\theta) \simeq \frac{(1-q)}{\theta}$ for $\theta \simeq 0$, in the first equation of (2.2) and obtain for $s \simeq 0$,

$$\dot\alpha \simeq H(p+q-1) - (p+q-1)\frac{G'(a)}{G(a)}\sin\alpha(0) + (q-1)\frac{\cot\alpha(0)}{s}.$$

Now integration gives

$$\alpha(s) - \alpha(0) \simeq C\, s + [(q-1)\cot\alpha(0)]\,\log s$$

for some $C \in \mathbf{R}$.

Clearly that is possible only if $\alpha(0) = \pi/2$; i.e. $\dot\eta(0) = 0$. //

(2.6) The previous lemma tells us that any solution γ which starts at the boundary $\theta = 0$ is orthogonal to it; we deduce that γ generates a smooth c.m.c. hypersurface, at least locally. Therefore a standard regularity theorem for elliptic P.D.E. insures that such a hypersurface is *analytic*, which in turn implies that $\alpha(s)$, $\eta(s)$, $\theta(s)$ are analytic functions of s around $s = 0$. This enables us to use the method of power series expansions.

(2.7) Any solution which starts at $\theta = 0$ can be expressed in the form $\eta = \eta(\theta)$ at least for small θ; since
$$ds^2 = G^2(\eta)\,d\theta^2 + d\eta^2 \,,$$
we obtain

(2.8) $\qquad \dfrac{d\theta}{ds} = \dfrac{1}{\sqrt{R^2(\theta) + G^2(\theta)}} \quad$ and $\quad \ddot\eta = \dfrac{1}{R^2(\theta) + G^2(\eta)}\dfrac{dR}{d\theta} \,,$

where

(2.9) $\qquad\qquad\qquad R(\theta) = \dfrac{d\eta}{d\theta}\,.$

Using (2.8), we transform (2.2)' into a differential equation for $\eta = \eta(\theta)$, to obtain
(2.10)
$$\begin{cases} R(\theta) = \dfrac{d\eta}{d\theta} \\ \theta\,\dfrac{dR}{d\theta} = -H m\, G(\eta)\theta\sqrt{R^2(\theta)+G^2(\eta)} + m\,\theta\,G'(\eta)G(\eta) + R(\theta)\,K(\theta)\,\theta \end{cases}$$

where $m = p + q - 1$, with initial conditions

(2.11) $\qquad\qquad \eta(0) = a, \quad \dfrac{d\eta}{d\theta}(0) = 0 = R(0)\,.$

(2.12) **Remark.** The function
$$F(\theta) = \begin{cases} K(\theta)\,\theta & \theta \neq 0 \\ -(q-1) & \theta = 0 \end{cases}$$

is analytic around $\theta = 0$.

The proof of Proposition (2.4) is based on the following technical

(2.13) Lemma. *Let*

$$\psi = \sum_{\substack{\ell+m+n+\nu \geq 2 \\ m+n+\nu \geq 1}} a_{\ell m n \nu} \, t^\ell \, \theta^m \, \tilde{\eta}^n \, R^\nu$$

be an analytic function of $(t, \theta, \tilde{\eta}, R)$ around $(0,0,0,0)$.

Consider the system

(2.14)
$$\begin{cases} \dfrac{\partial \tilde{\eta}}{\partial \theta} = R \\ \theta \dfrac{\partial R}{\partial \theta} = \lambda R + c\theta + \psi(t, \theta, \tilde{\eta}, R) \end{cases}$$

where $c, \lambda \in \mathbb{R}$ and λ is not a positive integer. Then there is a unique analytic solution $\tilde{\eta} = \tilde{\eta}(t, \theta)$ in a neighbourhood of $(0,0)$ which satisfies

(2.15) $\tilde{\eta}(t, 0) = 0, \quad \dfrac{\partial \tilde{\eta}}{\partial \theta}(t, 0) = R(t, 0) = 0$.

(2.16) Remark. It is implicit in the definition of ψ that for $\ell \geq 0$ we have assumed $a_{\ell 000} = a_{0001} = a_{0010} = a_{0100} = 0$.

The proof of Lemma (2.13) is given in (2.23) below.

Proof of Proposition (2.4). First we observe that (2.10) is of the form

(2.17)
$$\begin{cases} R(\theta) = \dfrac{d\eta}{d\theta} \\ \theta \dfrac{dR}{d\theta} = \Phi(\theta, \eta, R) \end{cases}$$

with Φ analytic around $(0, a, 0)$.

Next, we set

(2.18)
$$\begin{cases} \tilde{\eta}(t, \theta) = \eta(t, \theta) - a - t \\ \dfrac{\partial \tilde{\eta}}{\partial \theta}(t, \theta) = R(t, \theta) \end{cases}$$

and consider (for a fixed)

(2.19)
$$\begin{cases} \dfrac{\partial \tilde{\eta}}{\partial \theta} = R \\ \theta \dfrac{\partial R}{\partial \theta} = \Phi(\theta, \tilde{\eta} + a + t, R) \end{cases}.$$

Assume for a moment that (2.19) satisfies the hypotheses of Lemma (2.13). Then it admits an analytic solution $\tilde{\eta}(t, \theta)$ with

$$\tilde{\eta}(t, 0) = R(t, 0) = 0 .$$

Let $\eta(t, \theta)$ be associated with $\tilde{\eta}(t, \theta)$ as in (2.18); then clearly $\eta(\theta) = \eta(0, \theta)$ is an analytic solution of (2.10) which satisfies the initial conditions (2.11).

Moreover, Lemma (2.13) guarantees uniqueness and analytic dependence on the initial position a, because $\tilde{\eta}(t, \theta)$ is also analytic as a function of t.

In order to complete the proof of Proposition (2.4) we have only to show that the system (2.19) satisfies the hypotheses of Lemma (2.13). In fact, let

$$P(t, \theta, \tilde{\eta}, R) = \Phi(\theta, \tilde{\eta} + a + t, R)$$
$$= - H\ m\ G(\tilde{\eta} + a + t) \left[R^2(\theta) + G^2(\tilde{\eta} + a + t)\right]^{1/2} \theta$$
$$+ m\ \theta\ G'(\tilde{\eta} + a + t)\ G(\tilde{\eta} + a + t) + R(\theta)\ F(\theta)$$

by (2.10).

Then

$$\frac{\partial^\ell P}{\partial t^\ell}(0, 0, 0, 0) = 0 \qquad \ell \geq 0$$

$$\frac{\partial P}{\partial R}(0, 0, 0, 0) = -(q-1) = \lambda\ , \quad \lambda\ \text{not a positive integer}$$

$$\frac{\partial P}{\partial \tilde{\eta}}(0, 0, 0, 0) = 0 .$$

From these facts we conclude that (2.19) is of type (2.14), as required. //

(2.20) Proposition (2.4) has been established for an oriented curve $\gamma(s)$ for $s \geq 0$; and gives the existence of a branch starting at each point of the locus $\theta = 0$. There is also an *incoming branch*, corresponding to $s < 0$. That is obtained by replacing (2.8) with

$$\frac{d\theta}{ds} = -\frac{1}{\sqrt{R^2(\theta) + G^2(\theta)}} .$$

In this case (2.10) takes the form
(2.10)'
$$\begin{cases} R(\theta) = \dfrac{d\eta}{d\theta} \\ \theta\ \dfrac{dR}{d\theta} = H\ m\ G(\eta)\ \theta \sqrt{R^2(\theta) + G^2(\eta)} + m\ \theta\ G'(\eta)\ G(\eta) + R(\theta)\ F(\theta) . \end{cases}$$

If $H \neq 0$, then the two branches of γ form a *cusp* at the point $(a, 0)$. If $H = 0$, then (2.10) and (2.10)' coincide; and so do the two branches of the cusp; we shall say that the curve *bounces back* at $(a, 0)$. By way of summary we have

(2.21) Corollary. *If $(a, 0)$ (or $(a, \pi/2)$) is a point on the singular boundary of (2.2), then either*

$H = 0$, *in which case the curve γ_a bounces back; or*

$H \neq 0$, *so the curve γ_a has a cusp, which depends analytically on a.*

(2.22) Proposition (2.4) gives local existence of solutions γ starting from the singular boundary $\theta = 0$ (and similarly for $\theta = \pi/2$). Furthermore, equation (2.2) satisfies the standard Lipschitz condition on any compact domain without boundary points. Therefore we can consider solutions γ as defined and continuous on all $\mathbb{R}$, bearing in mind that when a solution hits the boundary, either it bounces back (case $H = 0$), or it passes from one branch of a cusp to the other (case $H \neq 0$).

(2.23) We proceed now to a proof of Lemma (2.13):

Step 1. We shall assume a formal power series solution of (2.14), compatible with (2.15); explicitly,

(2.24) $$\tilde{\eta}(t, \theta) = \sum_{\substack{i \geq 0 \\ j \geq 1}} b_{ij+1} t^i \theta^{j+1}, \quad R(t, \theta) = \sum_{\substack{i \geq 0 \\ j \geq 1}} b'_{ij} t^i \theta^j .$$

It follows from the first equation in (2.14) that

(2.25) $$b_{ij+1} = b'_{ij}/(j+1) \quad \text{for} \quad i \geq 0, j \geq 1.$$

And substituting (2.24) into the second equation in (2.14) gives
(2.26)
$$\sum_{\substack{i \geq 0 \\ j \geq 1}} (j - \lambda) b'_{ij} t^i \theta^j =$$

$$c\theta + \sum_{\substack{\ell+m+n+\nu \geq 2 \\ m+n+\nu \geq 1}} a_{\ell m n \nu} t^\ell \theta^m \left(\sum_{\substack{\alpha \geq 0 \\ \beta \geq 1}} b_{\alpha\beta+1} t^\alpha \theta^{\beta+1} \right)^n \left(\sum_{\substack{\gamma \geq 0 \\ \delta \geq 1}} b'_{\gamma\delta} t^\gamma \theta^\delta \right)^\nu .$$

By comparing coefficients,

(2.27) $$(j - \lambda) b'_{ij} = Q_{ij}(a_{\ell m n \nu}, b_{\alpha\beta}, b'_{\gamma\delta})$$

where Q_{ij} are polynomials with *nonnegative* integral coefficients for

(2.28) $\quad\quad \alpha, \gamma \leq i, \quad \beta, \delta \leq j, \quad \text{and} \quad \alpha + \beta, \gamma + \delta < i + j .$

(The last inequality would not generally hold without the assumption $a_{0010} = 0 = a_{0001}$.) It follows from (2.28) and the assumption $\lambda \neq j$ for all j that the b'_{ij} are uniquely determined, and so are b_{ij+1}. This proves uniqueness.

For instance,

(2.29)
$$(1-\lambda)b'_{01} = c$$
$$(1-\lambda)b'_{11} = a_{1100} + a_{1001}\, b'_{01}\, .$$

Step 2. We shall show that the formal solution (2.24) converges in a neighbourhood of $(0,0)$.

Since λ is not a positive integer, we can find a $K > 0$ such that

(2.30) $\qquad |j - \lambda| \geq 1/K \quad \text{for all} \quad j \geq 1\,.$

Also set

(2.31)
$$\begin{cases} C = |c| \\ A_{\ell m n \nu} = |a_{\ell m n \nu}| \\ \Phi(t, \theta, \tilde{\eta}, R) = \sum_{\substack{\ell+m+n+\nu \geq 2 \\ m+n+\nu \geq 1}} A_{\ell m n \nu}\, t^\ell\, \theta^m\, \tilde{\eta}^n\, R^\nu\,. \end{cases}$$

Note that Φ converges in a neighbourhood of $(0,0,0,0)$, because ψ does.

For purposes of comparison, define

(2.32)
$$\begin{cases} F_1(t, \theta, \tilde{\eta}, R) = \tilde{\eta} - \theta R \\ F_2(t, \theta, \tilde{\eta}, R) = -\frac{R}{K} + C\theta + \Phi(t, \theta, \tilde{\eta}, R)\,. \end{cases}$$

Then $F_1(0,0,0,0) = 0 = F_2(0,0,0,0)$; and

$$\det \begin{pmatrix} \partial F_1/\partial \tilde{\eta} & \partial F_1/\partial R \\ \partial F_2/\partial \tilde{\eta} & \partial F_2/\partial R \end{pmatrix}_{(0,0,0,0)} =$$

$$\det \begin{pmatrix} 1 & 0 \\ A_{0010} & -1/K + A_{0001} \end{pmatrix} = -1/K \neq 0\,.$$

Thus we can apply the implicit function theorem to conclude that the system

$$F_1(t, \theta, \tilde{\eta}, R) = 0$$
$$F_2(t, \theta, \tilde{\eta}, R) = 0$$

defines $\tilde{\eta}$, R as analytic functions of t, θ near $(0,0)$, with $\tilde{\eta}(0,0) = 0$, $R(0,0) = 0$.

Step 3. Because $A_{\ell,0,0,0} = 0 (\ell \geq 0)$, we can use (2.32) to express $\tilde{\eta}$ and R near $(0,0)$ as convergent power series of the form

(2.33)
$$\begin{cases} \tilde{\eta} = \sum\limits_{\substack{i \geq 0 \\ j \geq 1}} B_{ij+1} \, t^i \, \theta^{j+1} \\ R = \sum\limits_{\substack{i \geq 0 \\ j \geq 1}} B'_{ij} \, t^i \, \theta^j \end{cases},$$

i.e., $B_{\ell 1} = 0$, $B'_{\ell 0} = 0$ for all $\ell \geq 0$.

From the definition of F_1 we find

(2.34) $$B_{ij+1} = B'_{ij} \quad (i \geq 0, j \geq 1);$$

and, by substituting (2.33) into F_2,

$$B'_{ij}/K = Q_{ij}(A_{\ell m n \nu}, B_{\alpha \beta}, B'_{\gamma \delta})$$

where Q_{ij} are the polynomials in (2.27).

Because their coefficients are nonnegative, we can use (2.30) to conclude that

(2.35) $$B'_{ij} \geq |b'_{ij}| \quad \text{for all} \quad i \geq 0, j \geq 1.$$

(The special values (2.29) can serve as guides.) Taking into account (2.25) and (2.34), that implies

(2.36) $$B_{ij+1} \geq |b_{ij+1}| \quad \text{for all} \quad i \geq 0, j \geq 1.$$

Finally, the convergence of the series (2.33) together with the estimates (2.35), (2.36) insure the convergence of (2.24) near $(0,0)$. That completes the proof of Lemma (2.13). //

3. ANALYSIS OF THE O.D.E. AND PROOF OF THE MAIN THEOREM

We embed the system (1.11) in the following 1–parameter family of systems ($k > 0$), homothetically equivalent to (1.11):

(3.1)$_k$
$$\begin{cases} \dot{\alpha} = \dfrac{-3k \cos k\eta}{\sin k\eta} \sin \alpha + \dfrac{2k \cot 2\theta}{\sin k\eta} \cos \alpha \\ \dot{\eta} = \cos \alpha \\ \dot{\theta} = \dfrac{k \sin \alpha}{\sin k\eta} \end{cases}.$$

That is the system associated with the $SO(2) \times SO(2)$-invariant cohomogeneity 1 minimal submanifolds in $S^4(1/k)$, with metric

$$h_k = \frac{\sin^2 k\eta}{k^2} d\theta^2 + d\eta^2 \quad (0 \leq \eta \leq \pi/k)$$

on the associated orbit space. The obvious radial homothety

$$\rho(k) : S^4(1/k) \to S^4(1)$$

is $SO(2) \times SO(2)$-equivariant, and maps minimal hypersurfaces of $S^4(1/k)$ to those of S^4. Thus the solutions of $(3.1)_k$ are preserved by $\rho(k)$; more precisely, from straightforward substitution and uniqueness we obtain

(3.2) **Lemma.** *Let $\gamma(k, a, s) = (\eta(k, a, s), \theta(k, a, s))$ denote the unique solution of $(3.1)_k$ with $0 < k \leq 1, 0 < a \leq \pi/2k$ and initial conditions $\eta(k, a, 0) = a, \theta(k, a, 0) = 0$. Then for $0 < k_1, k_2 \leq 1$ we have*

(3.3)
$$\begin{cases} \frac{1}{k_1} \eta(k_2, a, s) = \eta(k_1 k_2, a/k_1, s/k_1) \\ \theta(k_2, a, s) = \theta(k_1 k_2, a/k_1, s/k_1) . \end{cases}$$

(3.4) As $k \to 0$ the system $(3.1)_k$ tends to

$(3.1)_0$
$$\begin{cases} \dot{\alpha} = -\dfrac{3 \sin \alpha}{\eta} + \dfrac{2 \cot 2\theta}{\eta} \cos \alpha \\ \dot{\eta} = \cos \alpha \\ \dot{\theta} = \dfrac{\sin \alpha}{\eta} . \end{cases}$$

Thus $(3.1)_0$ corresponds to the $SO(2) \times SO(2)$-action on $\mathbb{R}^4 = \mathbb{R}^2 \times \mathbb{R}^2$ and has the flat metric $h_0 = \eta^2 d\theta^2 + d\eta^2$ $(0 \leq \eta < +\infty)$ on its associated orbit space. Intuitively, the geometry of this action is a linear approximation to that on S^4 near the two poles. In particular, solutions of $(3.1)_0$ approximate those of $(3.1)_1 = (1.11)$ near the poles; i.e., for η near 0 or π. An advantage of $(3.1)_0$ is derived from its homothetic invariance; that permits us to reduce it to a vector field in the plane. The main result concerning $(3.1)_0$ is

(3.5) **Proposition.** *Let $a > 0$; then for $s \geq 0$, let*

$$\gamma(0, a, s) = (\eta(0, a, s), \theta(0, a, s))$$

be the unique solution of $(3.1)_0$ which satisfies $\eta(0, a, 0) = a$, $\theta(0, a, 0) = 0$. Then the function $\eta(0, a, s)$ is monotone increasing; and $\theta(0, a, s)$ oscillates indefinitely in a neighbourhood of $\theta = \pi/4$ with amplitude decreasing to 0:

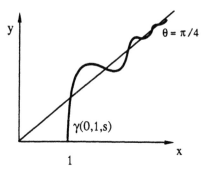

Figure 3.6

Remark. Proposition (2.4) insures local existence and uniqueness of $\gamma(0, a, s)$; in fact, $(3.1)_0$ is (2.2) with $G(\eta) = \eta$, $H = 0$, $p = q = 2$.

Now for the proof of (3.5):

Step 1. Let β denote the angle between γ and the x-axis: thus $\beta = \alpha + \theta$.

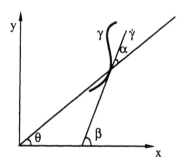

Figure 3.7

To express $(3.1)_0$ in terms of θ, β we multiply its first equation by $ds/d\theta = \eta/\sin\alpha$ to get

$$\frac{d\alpha}{d\theta} = -3 + 2\cot 2\theta \cot\alpha : \quad \text{i.e.,}$$

(3.8) $\qquad [2\cos(\beta + \theta)]d\theta = [\sin 2\theta \sin(\beta - \theta)]d\beta$.

Clearly the solutions of (3.8) are the trajectories of the vector field $X(\theta, \beta)$ given by

(3.9) $$\begin{cases} \dot\theta = \sin 2\theta \sin(\beta - \theta) \\ \dot\beta = 2\cos(\beta + \theta) \end{cases}$$

which we examine in the phase space $\{(\theta,\beta) : 0 \le \theta \le \pi/2, \beta \in \mathbf{R}\}$.

Step 2. Let R denote the compact region of (θ,β) space bounded by the lines

$$\theta = 0, \quad \theta = \pi/2, \quad \beta = \theta + \pi/2, \quad \beta = \theta - \pi/2.$$

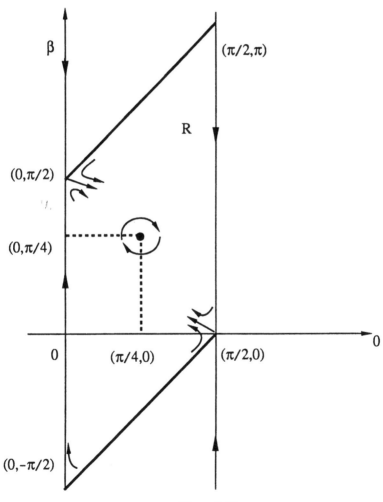

Figure 3.10

We observe that on the lines $\beta = \theta + \pi/2$ and $\beta = \theta - \pi/2$ the vector field X points into R. Moreover, by Lemma (2.5) no trajectory meeting Int R can touch either interval $\{(0, \beta) : -\pi/2 < \beta < \pi/2\}$ or $\{(\pi/2, \beta) : 0 < \beta < \pi\}$. We conclude that *any trajectory meeting R is thereafter trapped in R.*

Step 3. The singular points of (3.9) are

(3.11) $\qquad (0, m\pi + \pi/2), \quad (\pi/2, m\pi), \quad (\pi/4, m\pi + \pi/4)$

for $m \in \mathbf{Z}$. In particular, we have

(3.12) $(\pi/4, \pi/4)$ *is a focal point.*

(3.13) $(0, \pi/2)$ *and* $(\pi/2, 0)$ *are saddle points, with entrance (resp. exit) lines* $\theta = 0$ *and* $\theta = \pi/2$ *(resp.* $\beta = -\theta/2 + \pi/2$ *and* $\beta = -\theta/2 + \pi/4$*).*

Proof of (3.12). Set $u = \theta - \pi/4$, $v = \beta - \pi/4$; then (3.9) becomes

(3.14) $\qquad \begin{cases} \dot{u} = a_{11}u + a_{12}v + o_1\left(\sqrt{u^2 + v^2}\right) \\ \dot{v} = a_{21}u + a_{22}v + o_2\left(\sqrt{u^2 + v^2}\right) \end{cases}$

where the matrix $A = (a_{ij})$ is

$$A = \begin{pmatrix} -1 & 1 \\ -2 & -2 \end{pmatrix}.$$

Therefore

$$(\text{Trace } A)^2 - 4 \det A < 0$$

which asserts that $(0, 0)$ is a focal point of (3.14).

The proof of (3.13) is similar.

Step 4. *The vector field X does not have periodic orbits in R.*

Proof. Inspection of (3.9) shows that in R,

(3.15) $\qquad \begin{cases} \dot{\theta} > 0 \text{ for } \beta > \theta \text{ and } \dot{\theta} < 0 \text{ for } \beta < \theta; \\ \dot{\beta} > 0 \text{ for } \cot\theta > \tan\beta \text{ and } \dot{\beta} < 0 \text{ for } \cot\theta < \tan\beta \end{cases}$

That suggests dividing the region R into four subregions $R_1, \ldots, R_4$, as indicated in Figure 3.16

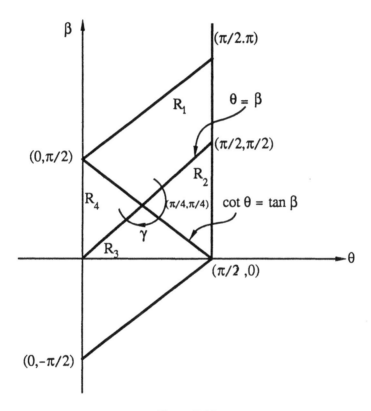

Figure 3.16

Suppose there were a periodic orbit γ. Then by a standard theorem [Ha, Thm. 3.1, p.150] it must enclose the stationary point $(\pi/4, \pi/4)$. Let $(\bar{\theta}, \bar{\beta})$ be a point of γ in R_1. Then from (3.15) we see that γ passes

(a) from R_1 to R_2 at a point (θ_1, β_1) with $\theta_1 = \beta_1$ and $\beta_1 < \bar{\beta}$;

(b) from R_2 to R_3 at (θ_2, β_2) with $\cot \theta_2 = \tan \beta_2$ and $\theta_2 < \theta_1$;

(c) from R_3 to R_4 at (θ_3, β_3) with $\theta_3 = \beta_3$ and $\beta_3 > \beta_2$;

(d) from R_4 to R_1 at (θ_4, β_4) with $\cot\theta_4 = \tan\beta_4$ and $\theta_4 > \theta_3$.

Taken together, these inequalities imply $\cot\beta_4 > \cot\beta_1$; so $\beta_4 < \beta_1$. On the other hand, $\beta_1 < \bar\beta$ and $\dot\beta < 0$ in R_1; which contradicts the fact that β increases in R_1 from β_4 to $\bar\beta$. We conclude that no periodic orbit γ exists in R.

Step 5. Taking Steps 2,3, and 4 together with the Poincaré– Bendixson theorem we obtain

(3.17) Lemma. *Every trajectory of (3.9) with at least one point in the interior of R spirals down to the focal point $(\pi/4, \pi/4)$. Thus it intersects the bisector $\theta = \beta$ at infinitely many distinct points $(t_n, t_n)_{n \geq 1}$ which converge to $(\pi/4, \pi/4)$ as $n \to +\infty$.*

Next, observe that the trajectories of the 1–parameter ($a > 0$) family of solutions $\gamma(0, a, s)$ correspond, in the (θ, β)-space, to a *unique* curve which starts at $(0, \pi/2)$ and points toward the interior of R. Therefore (3.17) applies; and Proposition (3.5) follows by reading (3.17) in (η, θ)-space. //

(3.18) Remark. The solutions $\gamma(0, a, s)$ are a homothetic family; i.e., with $a > 0$,

$$\eta(0, a, s) = a\, \eta(0, 1, s/a), \quad \theta(0, a, s) = \theta(0, 1, s/a).$$

(3.19) In the notations of Figure 1.3 and Lemma (3.2), we have

(3.20) Lemma. *For every integer $n > 0$ there is a sufficiently small $a > 0$ and a solution $\gamma(1, a, s)$ of $(3.1)_1$ which intersects the semi–axis O_1C at least n times.*

Proof. By Proposition (3.5), $\gamma(0, 1, s)$ intersects the θ-bisector infinitely often. Therefore, a sufficiently large segment $[0, \eta_0]$ of the θ-bisector intersects $\gamma(0, 1, s)$ at least n times.

We observe that, for $0 < a_0 < \eta, 0 \leq k$, the metrics h_k are analytic in both k and η; thus we can appeal to Proposition (2.4) to conclude that $\gamma(k, a, s)$ is analytic in k and a, for $0 \leq k \leq 1, a_0 < a$; thus solutions of $(3.1)_0$ approximate those of $(3.1)_k$, k small. In particular, for sufficiently small $k > 0$, with $\pi/2k > \eta_0$, the solution $\gamma(k, 1, s)$ approximates $\gamma(0, 1, s)$ so well that it intersects the segment $[0, \eta_0]$ (in the corresponding orbit space) in at least n points.

Then by the homothety Lemma (3.2) (with $k_1 = k = a, k_2 = 1$), $\gamma(1, a, s)$ $(a = k)$ is a solution of $(3.1)_1$ which intersects O_1C at least n times. //

From now on we restrict our attention to $(3.1)_1$ (=(1.11)).

(3.21) **Lemma.** *Let $\gamma = (\eta(s), \theta(s))$ be a solution of $(3.1)_1$. If $\gamma(s_0)$ is a point for which $0 < \eta(s_0) < \pi/2$ and $\dot\eta(s_0) = 0$, then $\ddot\eta(s_0) > 0$.*

Proof. Firstly, suppose $0 < \theta(s_0) < \pi/2$. Then the first equation in $(3.1)_1$ gives

$$\dot\alpha(s_0) = -3\cos\eta(s_0)\,\dot\theta(s_0).$$

Moreover, $\dot\theta(s_0) \neq 0$ because s is the arc length parameter; we conclude that $\dot\alpha(s_0)$ and $\dot\theta(s_0)$ are different from zero and of opposite sign. Now the conclusion follows immediately by

$$\ddot\eta = -\dot\theta\,\dot\alpha \sin\eta,$$

which can be easily derived from the second and third equations of $(3.1)_1$.

In case $\theta(s_0) = 0$ (and similarly if $\theta(s_0) = \pi/2$) consider the curve $\eta \equiv \eta(s_0)$; its pre-image in S^4 is a latitude hypersphere, with constant mean curvature $\neq 0$. Now observe that the pre-image of γ in S^4 has zero mean curvature; consequently, as s starts to increase beyond s_0 the curve γ must stay between $\eta \equiv \eta(s_0)$ and the minimal equator $\eta \equiv \pi/2$. Thus $\eta(s)$ increases after s_0, as asserted. //

(3.22) Let $\gamma_a(s) = (\eta_a(s), \theta_a(s))$ denote the solution $\gamma(1, a, s)$ of $(3.1)_1$; so $\eta_a(0) = a$, $\theta_a(0) = 0$.

(3.23) **Lemma.** *Suppose $0 < a < \pi/2$. Then $\eta_a(s)$ is monotone increasing for $\eta_a(s) \leq \pi/2$. And there exists $\bar s$ for which $\eta_a(\bar s) = \pi/2$ and $0 < \theta_a(\bar s) < \pi/2$.*

Proof. That η_a is increasing for $\eta(s) < \pi/2$ follows immediately from Lemma (3.21). Also, $\gamma_a(s)$ does not hit the boundary when $\eta_a(s) < \pi/2$; otherwise it would bounce back, as seen in (2.21). Therefore, because $\gamma(s)$ has infinite length, it is easy to deduce the existence of $\bar s$ such that $\eta_a(\bar s) = \pi/2$. And neither $\dot\eta_a(\bar s) = 0$ nor $\theta_a(\bar s) = 0$ (or $\theta_a(\bar s) = \pi/2$) is possible; for otherwise, $\eta_a(s) \equiv \pi/2$, by uniqueness. //

Finally we come to the proof of Theorem (1.7):

As a varies in $(0, \pi/2]$ we study the variation of intersections of γ_a with O_1C. It is convenient to restrict each γ_a to $[0, \ell(a)]$, where $\ell(a)$ is determined by

$$\eta_a(s) < \pi/2 \quad \text{for} \quad s < \ell(a); \quad \text{and} \quad \eta_a(\ell(a)) = \pi/2.$$

Lemma (3.23) insures that $\ell(a)$ is well defined and that $\gamma_a(s)$ does not hit the boundary for $0 < s \leq \ell(a)$. Moreover, γ_a depends analytically on a, by Proposition (2.4).

The basic idea of the construction is to show (by continuity arguments) the existence of a sequence

$$a_1 = \pi/2, \quad a_3, \quad a_5, \ldots$$

such that

(3.24) $\gamma_{a_{2i+1}}$ *intersects* O_1C *in exactly* $i + 1$ *points; and* $\gamma_{a_{2i+1}}(\ell(a_{2i+1})) = C$.

The Main Theorem follows using central symmetry; the required solutions are obtained from $\gamma_{a_{2i+1}}$ by reflection across C.

To construct (a_{2i+1}), we let $\sharp a$ = the number of intersections of γ_a and O_1C. Observe that $\gamma_{\pi/2}$ is the η–bisector, so $\sharp(\pi/2) = 1$. Next, we know from Lemma (3.20) that $\sharp a \to +\infty$ as $a \to 0$. We also observe that γ_a intersects O_1C transversally; for a tangential intersection would contradict uniqueness (because $\theta \equiv \pi/4$ is a solution of $(3.1)_1$).

For $i \geq 1$ we define

$$a_{2i+1} = \sup\{0 < a < \pi/2 : \sharp(a) \geq i + 1\}.$$

Then $\sharp(a_{2i+1}) = i + 1$, as we easily see from the transversality of intersections and the continuity of γ_a with respect to a. Analogously, $\gamma_{a_{2i+1}}(\ell(a_{2i+1})) = C$, proving (3.24), and hence the Main Theorem. //

4. Notes and comments

(4.1) The main ideas of the proofs in Section 2 were given in [HH]. A somewhat different development in the smooth context was presented in [FK].

(4.2) Proposition (3.5) is due to Bombieri, De Giorgi, and Giusti [BDG]. Step 4 was carried out in detail by Charlet [Cha]. In dimensions $n \geq 8$ the singularity (3.12) is of nodal type; that restricts the use of the standard representation of $SO(p) \times SO(q)$ in $SO(p + q + 1)$ to the case $p + q \leq 7$. The methods of this Chapter have been adapted [H2, II] to prove the existence of non–equatorial minimal embeddings of the $(n - 1)$–sphere in S^n for $n = 5, 6, 7$. And similarly [To] for $n = 2m$ $(m \geq 2)$, using the tensor representation of $SO(2) \otimes SO(m)$ in $SO(2m+1)$. It is unknown whether such embeddings exist for $n = 2m + 1$ $(m \geq 4)$. On the other hand, equivariant differential geometric methods can be used to construct non–equatorial minimal *immersions* of the $(n - 1)$–sphere in S^n for all $n \geq 4$ [HT].

(4.3) The singularity type (3.12) is related to the stability properties of the associated minimal cones in $\mathbb{R}^n$. That study is carried out in [HS].

CHAPTER VII. CONSTANT MEAN CURVATURE IMMERSIONS OF HYPERSPHERES IN $\mathbb{R}^n$

INTRODUCTION

In the spirit of Chapter VI, we present now some constructions of W–Y. Hsiang and collaborators. For each decomposition $n = p + q$ with $p, q \geq 2$, we obtain a countably infinite family of geometrically distinct $SO(p) \times SO(q)$–invariant immersions of the $(n - 1)$–sphere in $\mathbb{R}^n$ with constant mean curvature. These show that

(1) Hopf's Theorem (III (2.10)) does not hold in $\mathbb{R}^n$ for $n \geq 4$; and

(2) the assumption of embedding in Alexandrov's Theorem (II (2.6)) cannot be weakened to immersion.

1. STATEMENT OF THE MAIN THEOREM

(1.1) Let $SO(p) \times SO(q)$ act on $\mathbb{R}^n = \mathbb{R}^p \times \mathbb{R}^q$ by rotations in each factor ($p, q \geq 2$). Clearly the orbit space $\mathbb{R}^n / SO(p) \times SO(q)$ can be identified with the set

$$Q = \{(x, y) \in \mathbb{R}^2 : x, y \geq 0\}$$

with flat metric. And the volume function (up to a constant) is

$$V = V(x, y) = x^{p-1} \cdot y^{q-1}.$$

(1.2) As usual, denote by $\sigma : \mathbb{R}^n \to Q$ the projection onto the orbit space; let $\gamma(s) = (x(s), y(s))$ be a curve in Q parametrized by arc length s. According to formula IV (4.11), $\sigma^{-1}(\gamma) \subset \mathbb{R}^n$ has constant mean curvature H iff

$$(p + q - 1)H = k(\gamma) - \frac{d}{d\nu} \log V.$$

Since Q is flat, this equation can be rewritten (as in V (3.4))

(1.3) $\qquad \dot{x}\ddot{y} - \dot{y}\ddot{x} = (p + q - 1)H + (p - 1)\dfrac{\dot{x}}{y} - (q - 1)\dfrac{\dot{y}}{x}.$

For our purposes, it suffices to consider the case $H = 1$; in fact, if $\gamma = (x(s), y(s))$ is a solution of (1.3) with $H = 1$, then $\gamma_k = \left(\frac{1}{k}x(ks), \frac{1}{k}y(ks)\right)$ is a solution of (1.3) with $H = k \neq 0$. Geometrically, that corresponds to a suitable homothetic transformation and choice of orientation. Therefore we study

(1.4) $\qquad \dot{x}\ddot{y} - \dot{y}\ddot{x} = (p + q - 1) + (p - 1)\dfrac{\dot{x}}{y} - (q - 1)\dfrac{\dot{y}}{x}.$

Equation (1.4) is singular on the semi-axes $x = 0$ and $y = 0$, the boundary of Q. Nonetheless, we have the following existence

(1.5) Proposition. *Fix $a > 0$. Then for each point $P = (a, 0)$ on the x-axis there exists a unique solution curve γ_a of (1.4) which passes through P and has a perpendicular cusp point at P. Moreover, each branch of such a cusp is analytic; and the curves γ_a depend analytically on the parameter a. The same holds for points on the y-axis.*

Proof. Under the transformation $x = \eta \cos\theta$, $y = \eta \sin\theta$, $0 \leq \eta < +\infty$, $0 \leq \theta \leq \pi/2$, equation (1.4) becomes VI (2.2)′ with $G(\eta) = \eta$, $H = 1$; therefore the proof follows from Proposition VI (2.4), together with Corollary VI (2.21). //

(1.6) Let $\alpha(s)$ denote the angle between the tangent to $\gamma(s)$ and the x-axis:

(1.7) $$\dot{x}(s) = \cos\alpha(s), \qquad \dot{y}(s) = \sin\alpha(s).$$

(This is not the α of VI (2.2), nor that of VI (2.3ii)!)

Then (1.4) is equivalent to

(1.8) $$\dot{\alpha} = (p + q - 1) + (p - 1)\frac{\cos\alpha}{y} - (q - 1)\frac{\sin\alpha}{x}.$$

(1.9) According to the observation VI (2.22) each solution $\gamma(s)$ is defined for all $s \in \mathbb{R}$. Furthermore, we shall show in Proposition (2.15) that $\gamma(s)$ tends asymptotically to the line $x = (q - 1)/(p + q - 1)$ as $s \to +\infty$; and to the line $y = (p - 1)/(p + q - 1)$ as $s \to -\infty$. In terms of the angle $\alpha(s)$, that implies

(1.10) $$\lim_{s \to +\infty} \alpha(s) = \pi/2, \qquad \lim_{s \to -\infty} \alpha(s) = \pi \pmod{2\pi}.$$

Therefore it is natural to define globally the direction function $\alpha_\gamma(s) = \alpha(s)$ on a given γ in such a way that $\alpha_\gamma(s)$ has a jump of $+\pi$ at each cusp point and is continuous elsewhere.

Now (1.10) implies the existence of an *integer* $w(\gamma)$ such that

(1.11) $$\Delta\alpha_\gamma = \lim_{s \to +\infty} \alpha_\gamma(s) - \lim_{s \to -\infty} \alpha_\gamma(s) = 2\pi w(\gamma) - \pi/2.$$

The number $w(\gamma)$ is called the *winding number* of γ.

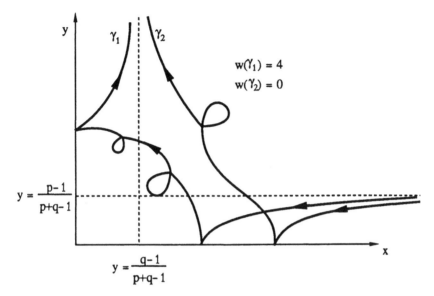

Figure 1.12

Roughly speaking, $w(\gamma)$ is the number of oriented loops of γ; and each cusp counts as one loop.

(1.12) We shall prove in Lemma (2.4) that each solution γ can have at most one cusp on each axis; therefore we classify solutions into the following types:

Type A. With no cusp point.

Type B. With exactly one cusp on the x-axis.

Type C. With exactly one cusp on the y-axis.

Type D. With exactly two cusps (one on each axis).

Type E. With one cusp at the origin.

(1.13) Let $z = (z_1, z_2)$ be a point of the orbit space Q; topologically, the pre-image of z under the projection $\sigma : \mathbb{R}^{p+1} \to Q$ is

$$\sigma^{-1}(z) = \begin{cases} S^{p-1} \times S^{q-1} & \text{if } z_1, z_2 > 0 \\ S^{p-1} & \text{if } z_1 = 0, z_2 > 0 \\ S^{q-1} & \text{if } z_1 > 0, z_2 = 0 \\ \text{a point} & \text{if } z = (0,0). \end{cases}$$

Therefore we can deduce the following properties:

(1.14) Take a curve γ of type D; and let $\gamma([s_0, s_1])$ be the part of γ between the two cusps. Then $\sigma^{-1}(\gamma[s_0, s_1])$ is an *immersed* S^{p+q-1} in $\mathbb{R}^{p+q}$; it is embedded if and only if γ does not have self–intersections on $[s_0, s_1]$.

(1.15) Similarly, the pre–images of curves (or segments of curves) of types A,B,C give rise to immersions of

$$\mathbb{R} \times S^{p-1} \times S^{q-1}, \quad \mathbb{R}^p \times S^{q-1}, \quad \mathbb{R}^q \times S^{p-1}$$

respectively in $\mathbb{R}^{p+q}$.

(1.16) **Theorem** (W–Y. Hsiang). (i) *If γ is a solution of type A, then $w(\gamma) \geq 0$. Conversely, for each integer $k \geq 0$, there exists a solution γ of type A with $w(\gamma) = k$.*

(ii) *If γ is a solution of type B (or C), then $w(\gamma) \geq 1$. Conversely, for each integer $k \geq 1$, there exists a solution γ of type B (or C) with $w(\gamma) = k$.*

iii) *If γ is a solution of type D, then $w(\gamma) \geq 2$. Conversely, for each integer $k \geq 2$, there exists a solution γ of type D with $w(\gamma) = k$.*

The proof is carried out in Sections 2, 3 below. As an immediate consequence, we have

(1.17) **Corollary.** *Fix $n = p + q, p, q \geq 2$. Then there exists a countably infinite family of non–congruent $SO(p) \times SO(q)$–invariant immersions of the $(n - 1)$–sphere in $\mathbb{R}^n$ of constant mean curvature 1.*

Proof. Apply (1.14) to iii) of (1.16). Only in the case $w(\gamma) = 2$ do we have an embedding which, by Alexandrov's Theorem, is the standard unit sphere. //

2. ANALYTICAL LEMMAS

(2.1) **Lemma.** *Let $[s_0, s_1]$ be any interval on which $\dot{\alpha} \equiv 0$. Then on $[s_0, s_1]$ either*

$$\alpha \equiv \pi \pmod{2\pi} \quad \text{and} \quad y \equiv \frac{p-1}{p+q-1}; \quad \text{or}$$

$$\alpha \equiv \pi/2 \pmod{2\pi} \quad \text{and} \quad x \equiv \frac{q-1}{p+q-1}.$$

Proof. Clearly the above two curves are solutions of (1.8). Conversely, let $\alpha \equiv \alpha_0$; then

$$0 = (p+q-1) + (p-1)\frac{\cos \alpha_0}{y} - (q-1)\frac{\sin \alpha_0}{x},$$

which defines a nondegenerate hyperbola if both $\cos \alpha_a$ and $\sin \alpha_0 \neq 0$. That contradicts $\dot\alpha \equiv 0$.

Also, neither $\alpha \equiv 0$ nor $\alpha \equiv 3/2\, \pi \pmod{2\pi}$ is acceptable, because $x, y \geq 0$. //

(2.2) These two functions play a basic role in this Section:

$$I = x^{q-1} \sin \alpha - \left(\frac{p+q-1}{q}\right) x^q; \quad J = y^{p-1} \cos \alpha + \left(\frac{p+q-1}{p}\right) y^p.$$

They are not conserved along solutions of (1.8); however, we observe the following monotonicity property, obtained simply by using (1.8):

(2.3) **Lemma.** *Along a solution γ of (1.8)*

$$\frac{dI}{ds} = (p-1) \frac{x^{q-1} \cos^2 \alpha}{y} \geq 0$$

$$\frac{dJ}{ds} = (q-1) \frac{y^{p-1} \sin^2 \alpha}{x} \geq 0.$$

In particular, both I and J are nondecreasing along γ.

As a first simple consequence, we obtain

(2.4) **Lemma.** *No solution γ of (1.8) can have more than one cusp on each axis.*

Proof. If $(0, b) = \gamma(s_0)$ is a cusp point on the y-axis, then $I(s_0) = 0$; and $I(s) > 0$ for $s > s_0$, by (2.3). Similarly for cusps on the x-axis. //

Now we establish some basic facts about the function I; the properties and proofs for J are similar.

(2.5) **Lemma.** *Along a solution of γ of (1.8) we have the bound*

$$I(s) \leq \frac{1}{q}\left(\frac{q-1}{p+q-1}\right)^{q-1} \quad (= \bar{I}, \text{ say})$$

with equality at some s_0 iff $x(s) = \frac{q-1}{p+q-1}$ for all $s \geq s_0$.

Proof. Set

(2.6) $$f(x) = x^{q-1} - \left(\frac{p+q-1}{q}\right) x^q.$$

Clearly $I(s) \leq f(x(s))$; and f assumes its maximum $\bar{I}$ at $\bar{x} = \frac{q-1}{p+q-1}$. Moreover, if $I(s_0) = f(\bar{x})$, then $\sin \alpha(s_0) = 1$; thus the conclusion follows by Lemma (2.1) and uniqueness. //

In the notation of (2.5), (2.6):

(2.7) Lemma. Let $0 < c < \bar{I}$; and let $m_c < M_c$ be the unique two points such that $f(m_c) = f(M_c) = c$. The solutions $(x(s), y(s))$ of

(2.8) $$I = x^{q-1} \dot{y} - \left(\frac{p+q-1}{q}\right) x^q = c$$

are a family of translationally invariant wave-like curves of period T oscillating between the lines $x = m_c$ and $x = M_c$. In particular, there is a positive constant $C(c)$ depending only on c such that for each solution of (2.8) there is an interval $[s_1, s_2]$ contained in a single period on which

(2.9) $$\dot{x}(s) \geq C(c).$$

Proof. The proof (for J instead of I) follows immediately from V (3.9). //

(2.10) Lemma. Let $\gamma = (x(s), y(s))$ be a solution of (1.8). Suppose that for some $s_0 \in \mathbb{R}$, $I(s_0) = c > 0$. Then

$$m_c \leq x(s) \leq M_c \quad \text{for} \quad s \geq s_0; \quad \text{and}$$

$$\dot{y} = \sin \alpha > \left(\frac{p+q-1}{q}\right) x \geq \left(\frac{p+q-1}{q}\right) m_c > 0.$$

Therefore $\lim_{s \to +\infty} y(s) = +\infty$.

Proof. $I(s) \geq I(s_0)$ for $s \geq s_0$, by Lemma (2.3). Thus $I(s) \leq f(x(s))$ implies $m_c \leq x(s) \leq M_c, s \geq s_0$. And $I > 0$ forces $\dot{y} > \left(\frac{p+q-1}{q}\right) x$. //

(2.11) **Lemma.** *Along a solution γ of (1.8)*

(2.12) $$\lim_{s \to +\infty} I(s) = \bar{I}.$$

Proof. Step 1. Lemmas (2.3) and (2.5) insure the existence of $\lim_{s \to +\infty} I(s) = c \leq \bar{I}$. We shall assume $0 < c < \bar{I}$, and derive a contradiction.

$\lim_{s \to +\infty} y(s) = +\infty$ by Lemma (2.10), so for large s the solution γ is C^1-approximated by a suitable solution β of

$$\dot{\alpha} = (p + q - 1) - (q - 1)\frac{\dot{y}}{x}.$$

This equation (with the roles of x and y interchanged) was thoroughly studied in V Section 3; in particular, see V (3.13); its solutions are uniquely determined by the prime integral

$$I = x^{q-1}\dot{y} - \left(\frac{p+q-1}{q}\right)x^q = a \quad \text{for} \quad a \in \mathbb{R}.$$

Because $\lim_{s \to +\infty} I(s) = c$, we see that γ can be C^1-approximated by a suitable solution β of $I = c$, for large s.

Step 2. Take s_1, s_2, T as in Lemma (2.7). For $m \in \mathbb{N}$ we estimate

(2.13)
$$\Delta I = I(s_2 + mT) - I(s_1 + mT) = \int_{s_1+mT}^{s_2+mT}\left(\frac{dI}{ds}\right)ds =$$
$$\int_{s_1+mT}^{s_2+mT}(p-1)\frac{x^{q-1}\dot{x}^2}{y}ds$$

With m sufficiently large, say $m > \bar{m}$, we can apply to γ the estimate that Lemma (2.7) provides for β; in particular, using (2.9) and the fact that $y(s)$ is increasing for s large, (2.13) give

(2.14) $$\Delta I \geq \frac{(p-1)C(c)}{y(s_2 + mT)}\int_{s_1}^{s_2} x^{q-1}\dot{x}\,ds.$$

Now choose $\tilde{T}$ such that $y(s_2) + m\tilde{T} = y(s_2 + mT)$; by the periodicity of β we can assume that $\tilde{T}$ is independent of $m > \bar{m}$.

If we substitute this in (2.14) we obtain

$$\Delta I \geq \frac{(p-1)\,d(c)}{y(s_2) + m\tilde{T}}$$

where $d(c)$ is a positive constant independent of $m > \tilde{m}$.

By summing over m we conclude that along γ

$$\lim_{s \to +\infty} I(s) \geq I(s_1 + \bar{m}\,T) + \sum_{m > \tilde{m}} \frac{d(c)}{y(s_2) + m\,\tilde{T}}.$$

That is a contradiction, for the series diverges, whereas I is bounded along γ.

A similar argument shows that also $\lim_{s \to +\infty} I(s) = c \leq 0$ is not possible. That completes the proof. //

We are now prepared to display the following qualitative behaviour of solutions at $\pm \infty$.

(2.15) Proposition. *If $\gamma(s)$ is a solution of (1.8), then*

$$x = \frac{q-1}{p+q-1} \quad \text{is the asymptotic line of } \gamma \text{ as } s \to +\infty; \text{ and}$$

$$y = \frac{p-1}{p+q-1} \quad \text{is the asymptotic line of } \gamma \text{ as } s \to -\infty.$$

Note that these two lines are the two trivial solutions of (1.8), as in Lemma (2.1).

Proof. From (2.12) we make an elementary algebraic calculation to see that

$$\lim_{s \to +\infty} \alpha(s) = \pi/2 \mod 2\pi\,;$$

and that $x = \frac{q-1}{p+q-1}$ is therefore the asymptote of γ as $s \to +\infty$.

Similar work with J proves the second statement. //

Now we proceed to study the geometry of solutions; namely, properties of the direction function $\alpha_\gamma = \alpha$.

(2.16) Lemma. *Suppose that along a solution $I(s_2) > 0$ (resp. $J(s_1) < 0$) for some $s_1, s_2 \in \mathbb{R}$. Then*

$$|\alpha(+\infty) - \alpha(s_2)| < \pi/2$$

$$(resp.\ |\alpha(-\infty) - \alpha(s_1)| < \pi/2)\,.$$

Proof. $I(s) \geq I(s_2)$ for $s \geq s_2$, by Lemma (2.3). Therefore $\sin \alpha(s) > 0$ for $s \geq s_2$, from which the first assertion follows. Similarly for J. //

As an immediate consequence of (2.16), we have

(2.17) Lemma. *Suppose $[s_1, s_2]$ is an interval for which $J(s_1) < 0$ and $I(s_2) > 0$. Then the winding number $w(\gamma)$ is uniquely determined by $\alpha(s_2) - \alpha(s_1)$; in fact, $w(\gamma)$ is the unique integer such that*

$$|2\pi w(\gamma) - \pi/2 - \alpha(s_2) + \alpha(s_1)| = |\alpha(+\infty) - \alpha(s_2) + \alpha(s_1) - \alpha(-\infty)| < \pi.$$

(2.18) Example 1. Let $\bar{\gamma}_1, \bar{\gamma}_2$ be the unique global solutions with single cusps at $\left(\frac{q-1}{p+q-1}, 0\right), \left(0, \frac{p-1}{p+q-1}\right)$ respectively. Then $w(\bar{\gamma}_1) = 1 = w(\bar{\gamma}_2)$.

Example 2. Take $0 < \delta < \min\left\{\frac{p}{(p+q-1)\sqrt{2}}, \frac{q}{(p+q-1)\sqrt{2}}\right\}$; and let γ be the unique global solution determined by $\alpha(0) = 3\pi/4$, $x(0) = \delta = y(0)$. Then $I(0) > 0$, $J(0) < 0$, so $w(\gamma) = 0$.

(2.19) Proposition. *For any positive integer k there is a global solution γ with $w(\gamma) \geq k$.*

Proof. Step 1. Because of the examples in (2.18) we can assume $k \geq 2$. Take $x_0, y_0 > 2(k+1)$, and let γ be the unique global solution with $x(0) = x_0, y(0) = y_0, \alpha(0) = 3\pi/4$. Let $s_1 < 0$ be the first point at which $y(s_1) = \sqrt{2}$; and $s_2 > 0$ the last at which $x(s_2) = \sqrt{2}$; then $\gamma([s_1, s_2])$ is the part of γ lying within the region $x \geq \sqrt{2}, y \geq \sqrt{2}$. We have

(2.20) $\qquad \dot{\alpha}(s) > M = \min\{p, q\} \quad \text{for all} \quad s \in [s_1, s_2]$.

Indeed,

$$|\dot{\alpha} - (p+q-1)| = \left|(p-1)\frac{\cos\alpha}{y} - (q-1)\frac{\sin\alpha}{x}\right|$$
$$< \{(p-1)|\cos\alpha| + (q-1)|\sin\alpha|\}/\sqrt{2} < \max\{(p-1), (q-1)\}.$$

Now (2.20) is immediate, so the difference

(2.21) $\qquad \Delta\alpha[s_1, s_2] = \alpha(s_2) - \alpha(s_1) = \int_{s_1}^{s_2} \dot{\alpha}\, ds > 4M\, k \geq 8k,$

because the lengths of both $\gamma([s_1, 0])$ and $\gamma([0, s_2])$ are $> 2k$.

Step 2. Along γ we have

(2.22)
$$\dot\alpha = \begin{cases} \dfrac{(p+q-1)}{q} + (p-1)\dfrac{\dot x}{y} - (q-1)\dfrac{I}{x^q} \\ \dfrac{(p+q-1)}{p} - (q-1)\dfrac{\dot y}{x} + (p-1)\dfrac{J}{y^p} \,. \end{cases}$$

We show that

(2.23) $$\Delta\alpha_\gamma = 2\pi w(\gamma) - \pi/2 \geq \Delta\alpha[s_1, s_2] - 2\pi\,.$$

If $J(s_1) < 0$ and $I(s_2) > 0$, then the inequality follows from Lemma (2.16). If $I(s) \leq 0$ and $\dot\alpha(s) \leq 0$, then $\dot x(s) < 0$ by (2.22). Therefore, if s_4 denotes the first time that $I = 0$, the angle $\alpha_{\gamma|_{[s_1, s_4]}}$ decreases by at most π.

The same conclusion holds on $[s_3, s_2]$ if s_3 denotes the first time that $J = 0$. Thus (2.23) follows by applying Lemma (2.16).

Step 3. Combining (2.21) with (2.23) gives

$$2\pi w(\gamma) - \pi/2 \geq 8k - 2\pi$$

which clearly implies $w(\gamma) \geq k$. //

3. PROOF OF THE MAIN THEOREM

(3.1) As further preparation for the proof of Theorem (1.16) we now perform certain deformations to reduce a solution with high winding number k to one with low winding number. As the deformation proceeds we obtain solutions with winding numbers $k - 1, k - 2, \ldots, 2, 1, 0$. Furthermore, we show that any decrease in winding number forces the occurrence of cusps. The main difficulty here is the qualitative behaviour of solutions obtained by small deformations of cusps.

(3.2) Let $\beta : [0, 1] \to \text{Int } Q$ be a curve (written $t \to \beta_t$) and v_t a unit vector field along β. The fundamental theorem of O.D.E. insures the existence of a unique family (γ_t) of global solutions of (1.8) such that

$$\gamma_t(0) = \beta_t, \qquad \dot\gamma_t(0) = v_t \quad (0 \leq t \leq 1)\,.$$

Such a family (γ_t) is said to be a *deformation of type 1*.

(3.3) A *deformation of type 2* is a family (γ^u) of global solutions of (1.8) with $\gamma^u(0) = (u, 0)$, where u varies in an interval $0 < x_1 \le u \le x_2$. (Equally well, we accept families of such solutions starting on the y–axis).

(3.4) A deformation of either type is *regular* if

i) whenever the curve $t \to \gamma_t(s_0)$ lies in Int Q, the function $\alpha_{\gamma_t(s_0)}$ is continuous in t;

ii) whenever $\gamma_0(s_0)$ is a cusp on the x– or y–axis, then so is $\gamma_t(s_0)$ for all t.

(3.5) **Lemma.** *Let (γ_t) be a deformation of type 1, and $[s_1, s_2]$ a finite interval containing 0. If*

$$\Gamma = \{\gamma_t(s) : s_1 \le s \le s_2, 0 \le t \le 1\} \subset \text{Int } Q,$$

then $\gamma_t|_{[s_1, s_2]}$ is regular.

Proof. Clearly there is a $c > 0$ for which $\Gamma \subset \{(x, y) : x \ge c, y \ge c\} = Q_c$. But (1.8) satisfies a Lipschitz condition on Q_c, so standard arguments apply. //

(3.6) **Lemma.** *Let (γ^u) be a deformation of type 2, and $[0, s_1]$ a finite interval. If*

$$\Gamma = \{\gamma^u(s) : 0 < s \le s_1, x_1 \le u \le x_2\} \subset \text{Int } Q,$$

then $\gamma^u|_{[0, s_1]}$ is regular.

Proof. Because (γ^u) depends analytically on u, there is a $\delta > 0$ such that $\beta(u) = \gamma^u(\delta)$ is an analytic curve on $[x_1, x_2]$ and $v_u = \dot\gamma^u(\delta)$ is an analytic vector field of unit length along β. Moreover, with any small fixed $\delta > 0$, the corresponding β has its image contained in a suitable region Q_c (as in (3.5)), for c small. Now Lemma (3.6) follows from Lemma (3.5). //

(3.7) **Lemma.** *If (γ_t) is a deformation of type 1 and if no γ_t meets the boundary, then $\Delta\alpha_{\gamma_t}$ is independent of t.*

Proof. The functions $I(s, t) = I(\gamma_t(s))$, $J(s, t) = J(\gamma_t(s))$ are both continuous. Consequently, there is an interval $[s_1, s_2]$ with $s_1 < 0 < s_2$ on which $I(s_2, t) >$

121

0, $J(s_1, t) < 0$ for all $0 \leq t \leq 1$. The claim now follows from Lemmas (2.17) and (3.5). //

Similarly, using Lemma (3.6) in place of Lemma (3.5), we obtain

(3.8) Lemma. *If (γ^u) is a deformation of type 2 and if no γ^u meets the boundary at any point other than $\gamma^u(0)$, then $\Delta \alpha_{\gamma^u}$ is independent of u.*

(3.9) Lemma. *If γ is a solution with no cusp points on the x–axis (resp., y–axis), then γ has at most one loop in the region $y < (p-1)/(p+q-1)$ (resp., $x < (q-1)/(p+q-1)$).*

Proof. Suppose γ has two loops in $y < (p-1)/(p+q-1)$. Then (see Figure below) there exists a point $\bar{s}$ such that $\cos \alpha(\bar{s}) = -1$ and $\dot{\alpha}(\bar{s}) \geq 0$:

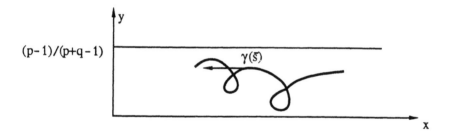

From (1.8) we have $\dot{\alpha}(\bar{s}) = (p+q-1) - (p-1)/y(\bar{s})$; but $y(\bar{s}) < (p-1)/(p+q-1)$, so $\dot{\alpha}(\bar{s}) < 0$ – a contradiction. Similarly for the alternative statement. //

(3.10) Lemma. *Let γ be a solution with a cusp at $\gamma(s_0) = (x_0, 0)$. Then there exists $\varepsilon > 0$ such that $\gamma([s_0 - \varepsilon, s_0])$ lies in the region $x \geq x_0$.*

Proof. Otherwise, for every $\varepsilon > 0$ there would be an $s_0 - \varepsilon \leq s < s_0$ with $\dot{\alpha}(s) < 0$, $\cos \alpha(s) > 0$, $\sin \alpha(s) < 0$; but that contradicts (1.8). //

We are now in position to prove the key

(3.11) Proposition. *Let (γ_t) be a deformation of type 1. Suppose that γ_0 is the only*

curve with a cusp point; and $\gamma_0(s_0)$ is its only cusp. Then either $w(\gamma_t)$ is constant
or $w(\gamma_t) = w(\gamma_0) - 1$ for $0 < t \leq 1$.

Proof. Lemma (3.7) shows that we need consider only small $t > 0$ – which amounts to showing that γ_t can have at most one loop near the cusp of γ_0. That was proved in Lemma (3.9). //

Similarly for

(3.12) **Proposition.** Let (γ^u) be a deformation of type 2. Suppose that γ^{x_1} is the only curve meeting the boundary twice. Then either $w(\gamma^u)$ is constant or $w(\gamma^u) = w(\gamma^{x_1}) - 1$ for $x_1 < u \leq x_2$.

We summarize pictorially the geometric possibilities arising in Proposition (3.11) during small deformations of γ_0:

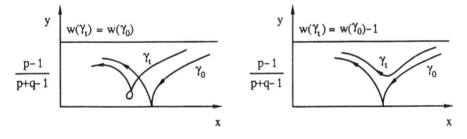

Figure 3.13

Proof of (1.16). Step 1. We have seen in Examples (2.18) that there are solutions of type A with winding number 0; and of type B and C with winding number 1. By Proposition (2.19), for any $k \geq 0$, there are global solutions γ with $w(\gamma) \geq k+2$.

Let (γ_t) be a deformation of type 1 with γ_0 of type A and $w(\gamma_0) = 0$; and $w(\gamma_1) \geq k + 2$.

By (3.7), (3.8), (3.11) and (3.12) the integer valued function $w(\gamma_t)$ can have jumps of absolute value at most 2; namely, as t increases, an increase (decrease) of 1 at the appearance (disappearance) of a cusp. Note that jumps of absolute value 2 can occur only when at some stage of the deformation curves of type A (resp., D) become of type D (resp., A).

Step 2. Now if γ_1 is of type A, as t decreases from 1, there is a t_0 for which $w(\gamma_{t_0}) \geq k + 2$ and γ_{t_0} has type B,C or D.

Figure 3.15 below illustrates the three typical stages which produce a decrease in $w(\gamma_t)$; namely, $t_1 > t_2 > t_3$ with $w(\gamma_{t_1}) = w(\gamma_{t_2}) = w(\gamma_{t_3}) + 1$.

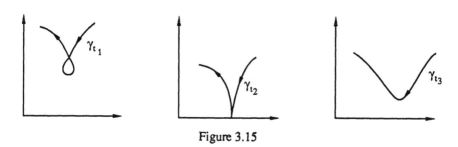

Figure 3.15

If γ_{t_0} is of type B (resp., C) we can deform it to the solution $\tilde{\gamma}_1$ (resp., $\tilde{\gamma}_2$) of Examples (2.18), by a deformation of type 2. As above, by Lemma (3.8) and Proposition (3.12), at some stage of that deformation we reach a solution of type D with winding number $\geq k + 2$.

Finally, if γ_{t_0} is of type D, a deformation of type 2 produces a solution of type B with winding number $\geq k + 1$.

By way of summary: *For any integer $k \geq 1$ there exist curves of types B,C,D with winding numbers $\geq k + 1$.*

Step 3. Start with one of these solutions of type D and deform it to a solution $\tilde{\gamma}_i$ of (2.18) by a deformation of type 2. Again by Lemma (3.8) and Proposition (3.12), at various stages of the deformation we meet solutions of type B,C,D whose winding numbers are all the integers less than the initial one.

Now let γ be a curve of type B and $w(\gamma) = k$, with a cusp at $(x_0, 0)$. Take a point (x_1, y_1) on γ near to $(x_0, 0)$, and let v_1 be a direction at (x_1, y_1) close to that of γ at (x_1, y_1). Then the solution starting at (x_1, y_1) in the direction v_1 is of type A and its winding number is k or $(k - 1)$. Thus we obtain a solution of type A with any prescribed winding number ≥ 0. On the other hand, $w(\gamma) \geq 0, 1, 2$ for a solution of type A, B or C, D respectively; for otherwise we violate the decreasing pattern pictured in Figure 3.15. That ends the proof of the main theorem. //

(3.16) If $p = q$, then (1.8) is symmetric with respect to the line $y = x$; that simplifies the study of deformations. In particular, we may study solutions determined by initial conditions $x(0) = a = y(0)$, $\alpha(0) = 3\pi/4$. These have winding number 0 for $a = a_0$ small (as in (2.18)); and winding number arbitrarily large for $a = a_1$ large (as in (2.19)). By symmetry, they are either of type A or D. When a ranges over $[a_0, a_1]$ we obtain solutions of types A and D with all intermediate *even* winding numbers:

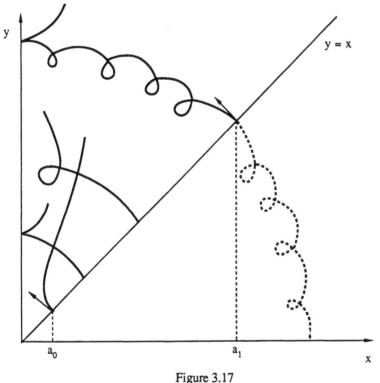

Figure 3.17

Similarly, we obtain the *odd* winding numbers by replacing $\alpha(0) = 3\pi/4$ with $\alpha(0) = 7\pi/4$.

(3.18) **Gauss maps.** Let $i_k : S^{p+q-1} \to \mathbb{R}^{p+q}$ be an immersion of constant mean curvature 1 associated with a solution $\gamma = \{(x(s), y(s)) : s \in [a, b]\}$ of type D with

$w(\gamma) = k+2$. Express S^{p+q-1} as a join $S^{p-1} * S^{q-1}$ as in IX (1.1). Then i_k induces the metric $g = i_k^*(\delta_{ij})$, which can be written via the join

$$g = x^2(s)g_{S^{p-1}} + y^2 g_{S^{q-1}} + ds^2 \quad \text{for} \quad s \in [a,b].$$

(3.19) The Gauss map Γ_k of i_k is

$$\Gamma_k : (S^{p+q-1}, g) \to S^{p+q-1}(1) \subset \mathbb{R}^p \times \mathbb{R}^q,$$

given by

(3.20) $\quad (s, u, v) \to (\dot{y}(s)u, -\dot{x}(s)\, v) = (\sin \alpha(s)u, -\cos \alpha(s)\, v)$

for $s \in [a,b], u \in S^{p-1}, v \in S^{q-1}$.

Γ_k *is a harmonic map* by Ruh–Vilms' Theorem (II (1.10)). $\Gamma_0 = id : S^{p+q-1}(1) \to S^{p+q-1}(1)$. More generally, the Brouwer degree of Γ_k is

(3.21) $\qquad \deg(\Gamma_k) = \begin{cases} 4k+1 & \text{if } p,q \text{ odd} \\ 1 & \text{all other cases.} \end{cases}$

Proof. Because γ has k loops in $[a, b]$, we have

$$\alpha(b) = \alpha(a) + \pi/2 + 2k\pi, \quad \alpha(a) = \pi/2 \bmod 2\pi.$$

Since we are in a purely homotopy context, we can assume

$$\alpha(s) = \pi/2 + \frac{(s-a)}{(b-a)}(2k\pi + \pi/2).$$

With this choice it is not difficult to count (with multiplicity) the points in the pre-image of a regular value and obtain (3.21). //

(3.22) **Remark.** Differentiating (1.8) gives

(3.23) $\quad \ddot{\alpha} + \left(\dfrac{(p-1)\dot{y}}{y} + \dfrac{(q-1)\dot{x}}{x}\right)\dot{\alpha} + \left(\dfrac{p-1}{y^2} - \dfrac{q-1}{x^2}\right)\sin\alpha\cos\alpha = 0.$

That is the Euler–Lagrange equation of the functional

(3.24) $\quad J(\alpha) = \displaystyle\int_a^b \left(\dot{\alpha}^2 + \dfrac{(p-1)\cos^2\alpha}{y^2} + \dfrac{(q-1)\sin^2\alpha}{x^2}\right) x^{q-1}\, y^{p-1}\, ds,$

subject to the conditions $\dot{x} = \cos\alpha, \dot{y} = \sin\alpha$.

We observe that (3.23) is the harmonicity equation of the Gauss map Γ_k in (3.19).

Such functionals J play a prominent role in our study (Chapters IX, X) of harmonic maps between spheres. We would be pleased if we could derive the existence theorems of the present Chapter by those variational methods.

4. Notes and comments

(4.1) Most of the results in this Chapter are due to W-Y. Hsiang [H4], following the special case ($p = q \geq 2$) – with a somewhat different proof – of [HTY].

(4.2) The main idea in Section 2 is to compare solutions of (1.4) with those of a more manageable one, which approximate (1.4) asymptotically. That sort of comparison has already been used in Chapter VI (3.20). And we shall apply it again in Chapters IX, X.

(4.3) If $H = 0$ in (1.3), the qualitative behaviour of the associated *minimal* hypersurfaces can be described by the methods of IV (3.5).

PART 3. HARMONIC MAPS BETWEEN SPHERES

CHAPTER VIII. POLYNOMIAL MAPS

INTRODUCTION

First we establish some basic properties of the Laplacian Δ on spheres. In particular, we compute its spectrum and eigenfunctions (Theorem (1.9)), in order to study *eigenmaps* (i.e., harmonic maps of constant energy density) $S^m \to S^n$. We characterize these as restrictions to S^m of maps $F : \mathbb{R}^{m+1} \to \mathbb{R}^{n+1}$ whose components are harmonic homogeneous polynomials of a common degree k (Corollary (1.10)).

Important examples of eigenmaps are

1) the Hopf fibrations $S^3 \to S^2, S^7 \to S^4, S^{15} \to S^8$, in the context of orthogonal multiplications (2.8);

2) the standard minimal immersions $S^m(c(k)) \to S^{n(k)}$, in a context of representation theory (1.17);

3) gradients of harmonic eiconals – a class of isoparametric functions (1.19).

In Section 2 we present some basic facts about orthogonal multiplications $\mathbb{R}^p \times \mathbb{R}^q \to \mathbb{R}^r$; in particular, from the Hopf construction we obtain a family of eigenmaps which includes the Hopf fibrations. We also describe a generalization to maps into Stiefel manifolds; and introduce bi-eigenmaps $S^p \times S^q \to S^r$.

The material of this Chapter provides a rich interplay involving algebra, geometry and topology; for the purposes of this monograph, eigenmaps, orthogonal multiplications and Hopf constructions are important as basic geometric tools for the construction of harmonic maps of spheres in the next two chapters.

At present, the classification of eigenmaps (and, more generally, of polynomial maps of spheres) seems far from reach; in Section 3 we prove theorems of R. Wood ((3.10), (3.16)) which impose restrictions on the dimensions m, n. Moreover, we display a close connection between the Hopf construction and quadratic maps (Theorem (3.19)).

1. EIGENMAPS $S^m \to S^n$

(1.1) Definition. A map $\varphi : S^m \to S^n$ is an *eigenmap with eigenvalue* λ if φ is harmonic and $|d\varphi|^2 \equiv \lambda$.

Let $i : S^n \to \mathbf{R}^{n+1}$ be the standard inclusion and set $\Phi = i \cdot \varphi$.

According to I (1.17) φ is an eigenmap with eigenvalue λ iff $\Delta\Phi = \lambda\Phi$. We proceed to derive explicit formulas for these.

(1.2) Lemma. *Let* $P : \mathbf{R}^{m+1} \to \mathbf{R}$ *be a function; set* $\tilde{P} = P|_{S^m}$. *Then*

$$(1.3) \qquad \Delta^{S^m} \tilde{P} = \left(\Delta^{\mathbf{R}^{m+1}} P + \frac{\partial^2 P}{\partial r^2} + m \frac{\partial P}{\partial r} \right) \circ i ,$$

where $i : S^m \to \mathbf{R}^{m+1}$ *is the standard inclusion, and* $\partial/\partial r$ *denotes radial differentiation.*

Proof. At any point $x \in S^m$ take an orthonormal base $(e_j)_{1 \leq j \leq m}$ of $T_x(S^m)$; and let $e_{m+1} = x$. Then

$$(-\Delta^{\mathbf{R}^{m+1}} P) \circ i = \sum_{j=1}^{m} \frac{\partial^2 P(x + te_j)}{\partial t^2}\bigg|_0 + \frac{\partial^2 P}{\partial r^2} \circ i$$

$$= \text{Trace } \nabla dP(di, di) + \frac{\partial^2 P}{\partial r^2} \circ i .$$

Also,

$$-\Delta^{S^m} \tilde{P} = dP(-\Delta^{S^m} i) + \text{Trace } \nabla dP(di, di) ,$$

so

$$-\Delta^{S^m} \tilde{P} = \left(-\Delta^{\mathbf{R}^{m+1}} P \right) \circ i - \frac{\partial^2 P}{\partial r^2} \circ i + dP(-\Delta^{S^m} i) .$$

Let $(\gamma_j)_{1 \leq j \leq m}$ denote geodesic arcs on S^m given by $\gamma_j(s) = \sin s \cdot e_j + \cos s \cdot x$, so

$$\gamma_j(0) = x, \gamma_j'(0) = e_j, \gamma_j''(0) = -x ;$$

$\Delta^{S^m} i = -\sum_{j=1}^{m} \gamma_j''(0)$, and we conclude that at x

$$dP(-\Delta^{S^m} i) = -m \frac{\partial P}{\partial r} \circ i . \;//$$

Let P be a k-homogeneous polynomial on $\mathbf{R}^{m+1}$; thus if $x \in \mathbf{R}^{m+1}$ and $r^2(x) = |x|^2$, then $P = r^k \tilde{P}$. Clearly

$$\frac{\partial P}{\partial r} = k\, r^{k-1}\, \tilde{P}, \quad \frac{\partial^2 P}{\partial r^2} = k(k-1) r^{k-2}\, \tilde{P}\,.$$

Therefore Lemma (1.2) implies

(1.4) Corollary. *If $P : \mathbf{R}^{m+1} \to \mathbf{R}$ is a harmonic k-homogeneous polynomial, then*

$$\Delta^{S^m} \tilde{P} = k(k+m-1)\tilde{P}\,.$$

I.e., $\tilde{P}$ is an eigenfunction of Δ^{S^m} with eigenvalue $\lambda_k = k(k+m-1)$.

We shall prove in Theorem (1.9) below that all eigenfunctions of Δ^{S^m} arise that way.

(1.5) We denote by $\mathcal{P}_k$ the vector space of k-homogeneous polynomials on $\mathbf{R}^{m+1}$; and note that $\mathcal{P}_k$ can be identified with the symmetric tensor power $\odot^k (\mathbf{R}^{m+1})^*$ of symmetric k-linear functions

$$\mathbf{R}^{m+1} \times \ldots \times \mathbf{R}^{m+1} \to \mathbf{R}\,.$$

A simple induction argument on m shows that

$$\dim \mathcal{P}_k = \binom{m+k}{k}\,.$$

Let $\mathcal{P} = \sum_{k \geq 0} \mathcal{P}_k$ denote the direct sum. We introduce an inner product on $\mathcal{P}$: For any $P, Q \in \mathcal{P}$ set

(1.6) $$<P, Q> = \int_{S^m} \tilde{P}\tilde{Q}\, dx\,.$$

Denote by

$$\mathcal{H}_k = \left\{ P \in \mathcal{P}_k : \Delta^{\mathbf{R}^{m+1}} P = 0 \right\}\,,$$

the vector space of harmonic k-homogeneous polynomials on $\mathbf{R}^{m+1}$; $\mathcal{H} = \sum_{k \geq 0} \mathcal{H}_k$.
And

$$\tilde{\mathcal{H}}_k = \left\{ \tilde{P} = P\big|_{S^m} : P \in \mathcal{H}_k \right\}\,.$$

(1.7) If $\mathcal{V}_\lambda$ denotes the eigenspace of $\Delta = \Delta^{S^m}$ associated with eigenvalue λ, then $\mathcal{V}_{\lambda_1} \perp \mathcal{V}_{\lambda_2}$ if $\lambda_1 \neq \lambda_2$; indeed, if $u_1 \in \mathcal{V}_{\lambda_1}$, $u_2 \in \mathcal{V}_{\lambda_2}$, then

$$\lambda_1 <u_1, u_2> \ = \ <\Delta u_1, u_2>$$
$$= <u_1, \Delta u_2> \ = \lambda_2 <u_1, u_2>,$$

from which the assertion follows.

(1.8) **Lemma.** *For every $k \geq 0$ we have the orthogonal decompositions*
$$\mathcal{P}_{2k} = \mathcal{H}_{2k} \oplus r^2 \mathcal{H}_{2k-2} \oplus \ldots \oplus r^{2k} \mathcal{H}_0;$$
$$\mathcal{P}_{2k+1} = \mathcal{H}_{2k+1} \oplus r^2 \mathcal{H}_{2k-1} \oplus \ldots \oplus r^{2k} \mathcal{H}_1.$$

Proof. Firstly, $\mathcal{P}_0 = \mathcal{H}_0 = \mathbb{R}$; and $\mathcal{P}_1 = \mathcal{H}_1$ is the space of linear forms on $\mathbb{R}^{m+1}$. We proceed by induction, assuming $\mathcal{P}_k = \mathcal{H}_k \oplus r^2 \mathcal{P}_{k-2}$ and proving $\mathcal{P}_{k+2} = \mathcal{H}_{k+2} \oplus r^2 \mathcal{P}_k$. Let us check that
$$\mathcal{H}_{k+2} + r^2 \mathcal{P}_k \subset \mathcal{P}_{k+2}$$
is an orthogonal direct sum: That is equivalent to showing $\widetilde{\mathcal{H}}_{k+2} \perp \widetilde{\mathcal{P}}_k$. We have seen in Corollary (1.4) that $\widetilde{\mathcal{H}}_{k+2} \subset \mathcal{V}_{\lambda_{k+2}}$, and by the induction hypothesis $\widetilde{\mathcal{P}}_k$ is contained in the sum of the other eigenspaces. These are mutally orthogonal, so $\widetilde{\mathcal{H}}_{k+2} \perp \widetilde{\mathcal{P}}_k$.

Finally, we must show that if $P \in \mathcal{P}_{k+2}$ is orthogonal to $\mathcal{P}_k$, then $\Delta P = 0$. Now $\Delta P \in \mathcal{P}_k$, and by induction it vanishes

iff $\Delta P \perp r^{2\ell} \mathcal{H}_{k-2\ell}$ for all $0 \leq 2\ell \leq k$;

iff $\widetilde{\Delta P}$ is orthogonal to all $\widetilde{\mathcal{H}}_{k-2\ell}$.

To verify that, take $H \in \mathcal{H}_{k-2\ell}$ and compute $\Delta \widetilde{PH} = \Delta \widetilde{P} \cdot \widetilde{H} + 2 <d\widetilde{P}, d\widetilde{H}> + \widetilde{P} \cdot \Delta \widetilde{H}$; thus we can decompose

$$0 = \int_{S^m} \Delta \widetilde{PH} \, dx = I + II + III = I + 3III.$$

But
$$III = \lambda_{k-2\ell} \int_{S^m} \widetilde{P} \, \widetilde{H} \, dx = 0$$
because $P \perp \mathcal{P}_k$. Also, because
$$\Delta \widetilde{P} = \widetilde{\Delta P} + (k+2)(m+k+1)\widetilde{P} \text{ by Lemma (1.2)},$$
$$0 = I = \int_{S^m} (\widetilde{\Delta P} \cdot \widetilde{H} + \lambda_{k+2} \widetilde{P} \cdot \widetilde{H}) dx$$
$$= \int_{S^m} \widetilde{\Delta P} \cdot \widetilde{H} \, dx. \ //$$

(1.9) Theorem. *The spectrum of Δ^{S^m} is*

$$\{\lambda_k = k(k+m-1) : k \in \mathbb{N}\} .$$

The eigenspace $V_{\lambda_k} = \widetilde{\mathcal{H}}_k$. Moreover,

$$\dim \widetilde{\mathcal{H}}_k = \binom{m+k}{k} - \binom{m+k-2}{k-2} = \frac{(2k+m-1)(k+m-2)!}{k!(m-1)!}$$

is a strictly increasing function of k for $m \geq 2$.

Proof. Lemma (1.8) shows that $\widetilde{\mathcal{H}} = \widetilde{\mathcal{P}}$ and that is dense in $C(S^m, \mathbb{R})$, by the Stone–Weierstrass theorem. Now every eigenvalue λ has the form $k(k+m-1)$; for otherwise its eigenspace $V_\lambda \perp \widetilde{\mathcal{P}}_k$ for all $k \geq 0$, contradicting the density of $\widetilde{\mathcal{P}}$ in $C(S^m, \mathbb{R})$.

Now $\widetilde{\mathcal{P}}_k = V_{\lambda_k}$; for otherwise, because $\widetilde{\mathcal{P}}_k$ is finite dimensional, there would exist $0 \neq Q \in V_{\lambda_k}$ orthogonal to $\widetilde{\mathcal{P}}_k$. But by construction, $Q \perp \widetilde{\mathcal{P}}_j$ for every $j \neq k$, again contradicting the density of $\widetilde{\mathcal{P}}$.

The statement about $\dim \widetilde{H}_k$ follows easily from the decomposition (1.8) and $\dim \mathcal{P}_k = \binom{m+k}{k}$. //

From (1.1) and Theorem (1.9) we obtain

(1.10) Corollary. *The eigenmaps $\varphi : S^m \to S^n$ are represented by $(n+1)$-tuples of harmonic k-homogeneous polynomials*

$$(\Phi^\alpha)_{1 \leq \alpha \leq n+1} \text{ in } \mathcal{H}_k \text{ for some } k \in \mathbb{N},$$

with

$$\sum_{\alpha=1}^{n+1} (\Phi^\alpha(x))^2 = 1 \text{ for all } x \in S^m .$$

(1.11) More generally, it is quite natural to study *eigenmaps* $M \to S^n$; namely, harmonic maps with constant energy density $\lambda/2$.

(1.12) Example. Let $\pi : M \to N$ be a harmonic Riemannian submersion, and $\varphi : N \to S^n$ a map. Then π is an eigenmap with $\lambda = n$. And as an application of IV (1.13), *φ is an eigenmap iff $\varphi \circ \pi$ is.*

(1.13) **Example.** If $\varphi_1 : M_1 \to S^{n_1}$ and $\varphi_2 : M_2 \to S^{n_2}$ are eigenmaps with the same eigenvalue, then so is

$$\varphi : M_1 \times M_2 \to S^{n_1+n_2+1}$$

where $\varphi(x_1, x_2) = \frac{1}{\sqrt{2}} (\varphi_1(x_1), \varphi_2(x_2))$.

Similarly for

$$\varphi_1 \otimes \varphi_2 : M_1 \times M_2 \to S^{n_1 n_2 + n_1 + n_2}.$$

(1.14) **Example.** A minimal Riemannian immersion $\varphi : M^m \to S^n$ is an eigenmap with $\lambda = m$. Indeed, $|d\varphi|^2 = \text{Trace}_g(\varphi^* g_{S^n}) = \text{Trace}_g g = m$.

Here is one way to construct such:

A *regular cone* in $\mathbf{R}^{n+1}$ is a subset C such that $C \setminus \{0\}$ is a manifold, and for every $x \in C$ and real number $a \geq 0$, $a \cdot x \in C$. For any submanifold $M \subset S^n$ the cone $C(M) = \{a \cdot x \in \mathbf{R}^{n+1} : a \geq 0, x \in M\}$ is regular. And if C is a regular cone, then $M = C \cap S^n$ is a submanifold. It is easy to verify that $M \subset S^n$ *is minimal iff* $C(M)/\{0\}$ *is minimal in* $\mathbf{R}^{n+1}$. For instance, we could take C holomorphic in $\mathbf{C}^{n+1}$: let $C = \{z \in \mathbf{C}^{n+1} : p_1(z) = \ldots = p_r(z) = 0\}$, where each p_j is a homogeneous polynomial; and suppose $C/\{0\}$ is a submanifold of $\mathbf{C}^{n+1}$.

(1.15) **Proposition.** *Let* $\Phi : M \to \mathbf{R}^{n+1}$ *be a Riemannian immersion such that* $\Delta \Phi = \lambda \Phi$, *for some* $\lambda \neq 0$. *Then*

a) $\lambda > 0$

b) $\Phi(M) \subset S^n(\sqrt{\frac{m}{\lambda}})$

c) $M \xrightarrow{\varphi} S^n(\sqrt{\frac{m}{\lambda}})$ *and the immersion* φ *is minimal.*
$\varphi \circ i = \Phi \searrow \quad \downarrow i$
$\qquad \qquad \mathbf{R}^{n+1}$

Proof. We observe that Φ is orthogonal to $\Phi(M)$, because $\Delta \Phi$ is. Thus, for any tangent vector field X on M

$$X < \Phi, \Phi > = 0$$

which implies $|\Phi|^2 = \text{constant} = r^2$. But

$$0 = \frac{1}{2} \Delta |\Phi|^2 = < \Phi, \Delta \Phi > - |d\Phi|^2 = \lambda r^2 - m,$$

and so $\lambda = m/r^2$. Now the minimality of φ follows immediately by I (1.17). //

Now, back to maps of spheres:

(1.16) In the notation of Theorem (1.9), set $n(k) = \dim \widetilde{H}_k - 1$ and $c(k) = \sqrt{\lambda_k/m}$. As an application of Proposition (1.15) we find

(1.17) **Corollary.** *Let* $(\Phi^1, \ldots, \Phi^{n(k)+1})$ *be an orthonormal base for* $\widetilde{H}_k$. *These (after normalization) are the components of a minimal Riemannian immersion*

$$\psi_{m,k} : S^m(c(k)) \to S^{n(k)}.$$

The $\psi_{m,k}$ are called the standard minimal immersions.

Proof. In order to apply Proposition (1.15), we have only to check that $\psi_{m,k}$ is a Riemannian immersion. Let $S^m = SO(m+1)/SO(m)$. The eigenspace $\widetilde{H}_k$ is $SO(m+1)$-invariant; it follows that the induced metric $\tilde{g} = \sum_{\alpha=1}^{n(k)+1} d\Phi^\alpha \odot d\Phi^\alpha$ is as well. Because $SO(m)$ acts *irreducibly* on the tangent space of S^m, we conclude that $\tilde{g} = c^2 h$, where h is the metric of $S^m(1)$ and c a positive constant. Then it is easy to see that $c = c(k)$. //

The standard map

$$\psi_{m,2} : S^m\left(\sqrt{\frac{2(m+1)}{m}}\right) \to S^{\frac{m(m+3)}{2}-1}$$

is sometimes called the (real) *Veronese map*. Note that $\psi_{m,2}$ is even, so that it factors to produce a minimal embedding of the real projective space

$$P^m\left(\sqrt{\frac{2(m+1)}{m}}\right) \hookrightarrow S^{\frac{m(m+3)}{2}-1}.$$

(1.18) We note that if we multiply the metric on $S^m(c(k))$ by the homothetic factor $(c(k))^{-2}$, we obtain an eigenmap $\varphi_{m,k} : S^m \to S^{n(k)}$ of polynomial degree k. These are *full*; i.e., their images are not contained in any totally geodesic hypersphere of $S^{n(k)}$.

(1.19) Next we describe a class of eigenmaps between spheres found by Reginald Wood [Wo1] – amongst Cartan's isoparametric functions.

A k–homogeneous polynomial $P : \mathbb{R}^{m+1} \to \mathbb{R}$ is an *eiconal* – in the language of geometric optics – if its gradient satisfies

(1.20) $\qquad |\nabla P(x)|^2 = |x|^{2k-2} \quad \text{for all} \quad x \in \mathbb{R}^{m+1}$.

Clearly $\varphi = \nabla P$ is a $(k-1)$–homogeneous polynomial map $\varphi : S^m \to S^m$. Furthermore, *if P is also harmonic, then φ is an eigenmap.* As in IV (3.5)

(1.21) **Proposition.** *For any regular value $t \in \mathbb{R}$ of a harmonic eiconal $P : \mathbb{R}^{m+1} \to \mathbb{R}$, the hypersurface $M_t = P^{-1}(t)$ has constant mean curvature. And the family (M_t) of such is parallel.*

(1.22) **Example.** Take $m = 1$, and use complex numbers $z = x_1 + i x_2$ on $\mathbb{R}^2 = \mathbb{C}$. Let $P_k : \mathbb{R}^2 \to \mathbb{R}$ denote the polynomial $P_k(z) = \operatorname{Re} z^k/k$. Then P_k is a harmonic eiconal; its gradient $\varphi = \nabla P_k : S^1 \to S^1$ is the eigenmap $\varphi_{k-1}(z) = \bar{z}^{k-1}$ with eigenvalue $\lambda_{k-1} = (k-1)^2$.

(1.23) For $m \geq 2$ a theorem of Münzner [Mün] asserts that harmonic eiconals exist only for degrees $k = 1, 2, 3, 4, 6$. The linear forms $P : \mathbb{R}^{m+1} \to \mathbb{R}$ are expressible as $P(x) = <a, x>$ for some $a \in S^m$. The quadratic forms $P : \mathbb{R}^{m+1} \to \mathbb{R}$ are those with signature $(\frac{m+1}{2}, \frac{m+1}{2})$; in particular, m must be odd.

When $k = 3$, they were classified by Cartan [C1]: For $\nu = 1, 2, 4, 8$ let $P : \mathbb{R}^{2+3\nu} \to \mathbb{R}$ be defined by

$$P = \frac{1}{3} u^3 - uv^2 + \frac{u}{2} \left(|X|^2 + |Y|^2 - 2|Z|^2 \right)$$
$$+ \frac{\sqrt{3} \, v}{2} \left(|X|^2 - |Y|^2 \right) + \frac{\sqrt{3}}{2} \left(XYZ + \bar{Z}\bar{Y}\bar{X} \right),$$

where $u, v \in \mathbb{R}$ and X, Y, Z are triples of real, complex, quaternion, or octonian numbers. Direct calculation shows that *P is a harmonic eiconal; its gradient $\varphi = \nabla P : S^m \to S^m$ is a quadratic eigenmap for $m = 4, 7, 13, 25$.*

All cubic harmonic eiconals arise in this way.

The Brouwer degree of φ is 2 when m is odd. And 0 when $m = 4$; however, in that case φ factors through the 2–fold covering *to give a harmonic representative of the nontrivial homotopy class of* $[\mathbb{R}P^4, S^4] = \mathbb{Z}_2$.

(1.24) When $k = 4$ there are only two harmonic eiconals, whose gradients are cubic eigenmaps $\varphi : S^m \to S^m$ for $m = 5, 9$ [C2].

Also when $k = 6$ there are only two harmonic eiconals, with $m = 7, 13$ ([C2], [Ab1]). The first gives an eigenmap $\varphi : S^7 \to S^7$ of polynomial degree 5 and Brouwer degree 1 ([EL4]).

(1.25) Finally, we define eigenmaps into the Stiefel manifolds $V_{n,k} = O(n)/O(n-k)$ of orthonormal k-frames in $\mathbf{R}^n$. We endow $V_{n,k}$ with the Riemannian metric induced by the standard embedding $i : V_{n,k} \to \mathbf{R}^n \times \ldots \times \mathbf{R}^n$ (k factors). That metric is clearly $O(n)$-invariant; however, it is not the canonical metric on $V_{n,k}$ as a homogeneous space [KN; II].

(1.26) With any map $\varphi : S^m \to V_{n,k}$ we denote its composition $\Phi = i \circ \varphi : S^m \to \mathbf{R}^n \times \ldots \times \mathbf{R}^n$.

We can regard Φ as an $(n \times k)$-matrix function with columns $\Phi_1, \ldots, \Phi_k$. Define also the $(k \times k)$-matrix function $D\Phi$ by

(1.27)
$$(D\Phi)_{ij} = \langle d\Phi_i, d\Phi_j \rangle$$

where

$$\langle d\Phi_i, d\Phi_j \rangle = \sum_{\gamma=1}^{n} g_{S^m}^{\alpha\beta} \frac{\partial \Phi_i^\gamma}{\partial x^\alpha} \frac{\partial \Phi_j^\gamma}{\partial x^\beta}.$$

(1.28) **Proposition.** *A map $\varphi : S^m \to V_{n,k}$ is harmonic iff*

(1.29)
$$\Delta^{S^m} \Phi = \Phi \cdot D\Phi.$$

Moreover $|d\varphi|^2 = |d\Phi|^2 = \operatorname{Trace} D\Phi$.

Proof. Because i is a Riemannian embedding, Corollary I (1.16) tells us that φ is harmonic iff $\Delta^{S^m} \Phi$ is everywhere orthogonal to $V_{n,k}$. In terms of Cartesian coordinates (x_j^i) on $\mathbf{R}^n \times \ldots \times \mathbf{R}^n$ ($1 \leq i \leq n, 1 \leq j \leq k$), $V_{n,k}$ is the zero set of the algebraic equations

$$\sum_{i=1}^{n} x_j^i x_h^i = \delta_{jh} \quad (1 \leq j, h \leq k).$$

Thus we compute the tangent space of $V_{n,k}$ at a point $A = (A_j^i)$ (translated to the origin)

$$T_A(V_{n,k}) = \left\{ v = (v_j^i) : \sum_{i=1}^{n} v_\alpha^i A_\beta^i + v_\beta^i A_\alpha^i = 0, \quad 1 \leq \alpha \leq \beta \leq k \right\}.$$

Now it is easy to see that the condition

$$\Delta^{S^m} \Phi(y) \in T^\perp_{\Phi(y)}(V_{n,k})$$

is expressible as

$$\Delta^{S^m} \Phi = \Phi \cdot D$$

for some symmetric $(k \times k)$–matrix function D. But

$$<\Phi_j, \Phi_i> = \delta_{ji}, \quad <\Delta^{S^m}\Phi_j, \Phi_i> = <d\Phi_j, d\Phi_i>,$$

so that

$$D_{ij} = <d\Phi_i, d\Phi_j> = (D\Phi)_{ij}, \quad \text{as required.}$$

The final assertion $|d\varphi|^2 = |d\Phi|^2 = \text{Trace } D\Phi$ is obvious. //

(1.30) We observe that if $k = 1$ then $V_{n,1} = S^{n-1}$; and Proposition (1.28) becomes Proposition I (1.17). It seems natural to call *eigenmaps* those solutions of (1.29) for which $(D\Phi)_{ii} = \lambda_i = $ constant ($1 \leq i \leq k$). These have constant energy density.

(1.31) **Example.** If $\varphi : S^r \to V_{n,k}$ is a totally geodesic embedding and $\psi : S^m \to S^r$ is an eigenmap with eigenvalue λ, then $\varphi \circ \psi : S^m \to V_{n,k}$ is an eigenmap with $\lambda_i = \lambda (1 \leq i \leq k)$.

2. ORTHOGONAL MULTIPLICATIONS AND RELATED CONSTRUCTIONS

(2.1) An *orthogonal multiplication* is a bilinear map $f : \mathbb{R}^p \times \mathbb{R}^q \to \mathbb{R}^n$ which is norm–preserving:

$$|f(x,y)| = |x| \, |y| \quad \text{for all} \quad x \in \mathbb{R}^p, y \in \mathbb{R}^q.$$

Clearly its restriction (still denoted by f) $S^{p-1} \times S^{q-1} \to S^{n-1}$ is an eigenmap with $\lambda = p + q - 2$; and a totally geodesic embedding in each variable separately.

(2.2) Remark. Take $n = q$. Then for any $x \in S^{p-1}$ the map $y \to f(x, y)$ is an element of $O(q)$; i.e., f induces a map $S^{p-1} \to O(q)$, which is a totally geodesic embedding.

(2.3) The *Hopf construction on f* is the map
$$F : \mathbf{R}^p \times \mathbf{R}^q \to \mathbf{R}^n \times \mathbf{R}^1 = \mathbf{R}^{n+1}$$
given by $F(x, y) = (2f(x, y), |x|^2 - |y|^2)$. Because $|F(x, y)|^2 = (|x|^2 + |y|^2)^2$, its restriction defines a map

(2.4) $$\varphi : S^{p+q-1} \to S^n ,$$

also called the Hopf construction on f. The components $(F^\alpha)_{1 \le \alpha \le n}$ are harmonic quadratic polynomials; and $\Delta F^{n+1} = 2p - 2q$. Therefore we obtain

(2.5) Lemma. *The Hopf construction (2.4) is an eigenmap iff $p = q$; the associated eigenvalue $\lambda = 4p$.*

(2.6) We shall write $f(p, q; n)$ for an orthogonal multiplication as in (2.1); and denote by $p * q$ the smallest integer n for which there is an $f(p, q; n)$.

Problems. Determine $p * q$. Classify the orthogonal multiplications $f(p, q; n)$. That was posed by Hurwitz (1898); its solution remains far from our reach. A survey and some new results have been given in [La2], [La3].

Here are some special cases of importance to us:

(2.7) Theorem (Hurwitz–Radon). *Write q in the form $q = (2c + 1)2^{a+4b}$, where $0 \le a \le 3$, and define $\rho(q) = 2^a + 8b$. Then there exists an orthogonal multiplication $f(p, q; q)$ iff $p \le \rho(q)$. In particular, there exists an $f(p, p; p)$ iff $p = 1, 2, 4, 8$; real, complex, quaternion, and octonian multiplications are examples. Also, $p * q \le q$ iff $p \le \rho(q)$.*

(2.8) Examples (Hopf's fibrations). Taking complex multiplication $f : \mathbf{R}^2 \times \mathbf{R}^2 \to \mathbf{R}^2$ and performing Hopf's construction produce the eigenmap $h = \varphi : S^3 \to S^2$ with eigenvalue $\lambda = 8$.

From quaternion multiplication $f : \mathbf{R}^4 \times \mathbf{R}^4 \to \mathbf{R}^4$ we obtain the eigenmap $\eta = \varphi : S^7 \to S^4$ with eigenvalue $\lambda = 16$.

And from octonian multiplication $f: \mathbb{R}^8 \times \mathbb{R}^8 \to \mathbb{R}^8$ we obtain the eigenmap $\sigma = \varphi: S^{15} \to S^8$ with eigenvalue $\lambda = 32$.

In all three cases the map φ is a Riemannian fibration, provided that the base sphere has radius $1/2$. The fibres are great spheres S^1, S^3, S^7 respectively. h is the generator of $\pi_3(S^2) = \mathbb{Z}$; η is the element $(1,0)$ of $\pi_7(S^4) = \mathbb{Z} \oplus \mathbb{Z}_{12}$; and σ the element $(1,0)$ of $\pi_{15}(S^8) = \mathbb{Z} \oplus \mathbb{Z}_{120}$.

(2.9) Lam [La3] has lower bounds on $p * q$. He has compiled the following table for $p * q$ for small values of p, q:

p\q	1	2	3	4	5	6	7	8	9	10	11	12	13	14	15	16	17
1	1	2	3	4	5	6	7	8	9	10	11	12	13	14	15	16	17
2		2	4	4	6	6	8	8	10	10	12	12	14	14	16	16	18
3			4	4	7	8	8	8	11	12	12	12	15	16	16	16	19
4				4	8	8	8	8	12	12	12	12	16	16	16	16	20
5					8	8	8	8	13	14	15	16	16	16	16	16	21
6						8	8	8	14	14	16	16	16	16	16	16	22
7							8	8	15	16	16	16	16	16	16	16	23
8								8	16	16	16	16	16	16	16	16	24
9									16	16	16	16	16	16	16	16	25
10										16							

(2.10) In view of Lemma (2.5) we have special interest in the orthogonal multiplications $f(p, p; n)$; and in $p * p$. We always have the tensor product multiplication $f(p, p; p^2)$, taking $\mathbb{R}^{p^2} = \mathbb{R}^p \otimes \mathbb{R}^p$ with its tensor norm. Therefore, in looking for $p * p$ we restrict our attention to $p \leq n \leq p^2$.

Lam [La1] found an $f(10, 10; 16)$; the Hopf construction on it represents twice the generator of $\pi_{19}(S^{16}) = \mathbb{Z}_{24}$. Adem [Ad] exhibited an $f(17, 17; 32)$ as restriction of an $f(17, 18; 32)$.

Again, there are significant lower bounds for $p * p$: for instance, $12 * 12 \geq 20$, $15 * 15 \geq 24$ [Y].

(2.11) A map $f: S^{p-1} \times S^{q-1} \to S^{n-1}$ is a *bi-eigenmap with bi-eigenvalue* (λ, μ) if for each fixed $y \in S^{q-1}$ (resp. $x \in S^{p-1}$) the map $x \to f(x, y)$ (resp. $y \to f(x, y)$) is an eigenmap with eigenvalue λ (resp., μ).

(2.12) Example. An orthogonal multiplication $f : S^{p-1} \times S^{q-1} \to S^{n-1}$ is a bi-eigenmap with bi-eigenvalue $(p-1, q-1)$. Let $\varphi_1 : S^u \to S^{p-1}, \varphi_2 : S^v \to S^{q-1}$ be eigenmaps with eigenvalues λ_1, λ_2 respectively. Then $f(\varphi_1, \varphi_2)$ is a bi-eigenmap $S^u \times S^v \to S^{n-1}$ with bi-eigenvalue (λ_1, λ_2). Similarly, $\varphi_1 \otimes \varphi_2 : S^u \times S^v \to S^{pq-1}$ is a bi-eigenmap with bi-eigenvalue (λ_1, λ_2).

(2.13) Example. Identifying S^3 with the unit quaternions and S^2 with the unit quaternions with zero real part, we define

$$f : S^2 \times S^3 \to S^2 \quad \text{by} \quad f(x, y) = y \cdot x \cdot \bar{y},$$

where $\cdot$ denotes quaternion multiplication. Thus f is a bi-eigenmap with bi-eigenvalue $(2, 8)$.

(2.14) With the Riemannian metric of (1.25) on a Stiefel manifold, an orthogonal multiplication $f : \mathbb{R}^p \times \mathbb{R}^q \to \mathbb{R}^n$ determines a totally geodesic embedding $\varphi_f : S^{p-1} \to V_{n,q}$ by

$$\varphi_f(x) = (f(x, e_1), \ldots, f(x, e_q)).$$

Thus φ_f is an eigenmap with $\lambda_i = p - 1$ $(1 \leq i \leq q)$.

Take $p = n = q$. Then $\varphi_f : S^{p-1} \to O(p)$ is the characteristic map of the tangent bundle of S^p.

(2.15) Proposition. *The construction in (2.14) determines a bijective correspondence between orthogonal multiplications $f : \mathbb{R}^p \times \mathbb{R}^q \to \mathbb{R}^n$ and totally geodesic embeddings $\varphi : S^{p-1} \to V_{n,q}$ satisfying $\varphi(-x_0) = -\varphi(x_0)$ for some point $x_0 \in S^{p-1}$.*

Proof. Step 1. We construct the inverse of the correspondence in (2.14). Suppose $\varphi : S^{p-1} \to V_{n,q}$ is a map such that for any orthonormal pair $u, v \in \mathbb{R}^p$,

$$<\varphi_i(u), \varphi_j(v)> + <\varphi_i(v), \varphi_j(u)> = 0.$$

Then φ defines an orthogonal multiplication $f(p, q; n)$.

Proof. Letting (u_i) and (e_j) denote the standard bases of $\mathbb{R}^p$ and $\mathbb{R}^q$ define $f : \mathbb{R}^p \times \mathbb{R}^q \to \mathbb{R}^n$ as the bilinear extension of $(u_i, e_j) = \varphi_j(u_i)$. Then $v \to f(u_i, v)$ is a map $S^{q-1} \to S^{n-1}$.

We are thereby reduced to showing that $f(u_i, v) \perp f(u_j, v)$ for $i \neq j$. But for $v = \Sigma v^s e_s$,

$$< f(u_i, v), f(u_j, v) > = \sum_s (v^s)^2 < \varphi_s(u_i), \varphi_s(u_j) >$$
$$+ \sum_{s \neq t} v^s v^t < \varphi_s(u_i), \varphi_t(u_j) > = 0$$

because each term in the first sum vanishes, and those of the second cancel in pairs.

Step 2. Given $\varphi : S^{p-1} \to V_{n,q}$ as in (2.15) and a geodesic γ in S^{p-1}, then each $\varphi_i \circ \gamma$ is a geodesic in S^{n-1}; and φ_i preserves arc length.

Proof. Let $x_0 \in S^{p-1}$ be such that $\varphi(-x_0) = -\varphi(x_0)$. If γ is a geodesic of unit velocity from x_0 to $-x_0$, then $\varphi \circ \gamma$ is a path from $\varphi(x_0)$ to $-\varphi(x_0)$ of length $q\pi$. That is the distance between $\varphi(x_0)$ and $-\varphi(x_0)$ in $V_{n,q}$; and can be achieved only by a path of length π on each S^{n-1}. Therefore each $\varphi_i \circ \gamma$ is a geodesic of S^{n-1} of unit length.

Repeating that argument following γ back to x_0, we find $\varphi(-x) = -\varphi(x)$ for all x on γ. In fact, that is true for all $x \in S^{p-1}$, because we can fill out S^{p-1} with such geodesics. That in turn allows us to repeat the argument for any geodesic in S^{p-1}, and obtain the desired conclusion.

Step 3. If γ_1, γ_2 are geodesics in S^{n-1} with unit velocity such that $< \gamma_1(t), \gamma_2(t) > \equiv 0$. Then

$$< \gamma_1(t), \gamma_2(t + \pi/2) > + < \gamma_1(t + \pi/2), \gamma_2(t) > \equiv 0.$$

Indeed, we can assume

$$\gamma_1(t) = \cos t \, e_1 + \sin t \, e_2, \quad \gamma_2(t) = \cos t \cdot v + \sin t \cdot w,$$

with

$$v = a \, e_2 + b \, e_3$$
$$w = -a \, e_1 + c \, e_4.$$

The assertion follows now by simple calculation.

Step 4. Finally, take orthogonal vectors $v, w \in S^{p-1}$; and γ a geodesic of unit velocity in their plane. Given i, j, let $\gamma_1 = \varphi_i \circ \gamma$ and $\gamma_2 = \varphi_j \circ \gamma$; these are geodesics on S^{n-1} with unit velocity, by Step 2 above. Thus the conditions in Step 1 are satisfied if $i = j$; and also if $i \neq j$, because γ_1 and γ_2 are orthogonal and so Step 3 applies. That completes the proof of Proposition (2.15). //

(2.16) We now construct eigenmaps into Stiefel manifolds, as in (1.30).

Let $f : \mathbf{R}^p \times \mathbf{R}^q \times \mathbf{R}^r \to \mathbf{R}^n$ be a trilinear map which is norm–preserving

$$|f(x,y,z)| = |x|\,|y|\,|z| \quad \text{for all} \quad (x,y,z) \in \mathbf{R}^p \times \mathbf{R}^q \times \mathbf{R}^r.$$

We define the *Hopf construction on f* as the map $F : \mathbf{R}^p \times \mathbf{R}^q \times \mathbf{R}^r \to \mathbf{R}^n \times \mathbf{R}^r$ given by

$$F(x,y,z) = (2f(x,y,z),\ (|x|^2 - |y|^2)z).$$

Its restriction defines a map

(2.17) $$\varphi : S^{p+q-1} \to V_{n+r,r},$$

where $\varphi(x,y) = (F(x,y,e_1),\ldots,F(x,y,e_r))$ for $(x,y) \in S^{p+q-1} \subset \mathbf{R}^p \times \mathbf{R}^q$. It follows from the definitions that $\varphi(x,y)$ is indeed an orthonormal r–frame in $\mathbf{R}^{n+r}$.

(2.18) **Proposition.** *If $p = q$, then φ in (2.17) is an eigenmap with eigenvalues $\lambda_i = 4p$ $(1 \leq i \leq r)$.*

Proof. Each component $\varphi_i : S^{2p-1} \to S^{n+r-1}$ is an eigenmap with $\lambda_i = 4p$, by Lemma (2.5) $(1 \leq i \leq r)$. //

(2.19) **Example.** Take an orthogonal multiplication $f_0 : \mathbf{R}^p \times \mathbf{R}^r \to \mathbf{R}^r$ with $p = \rho(r)$, as in Theorem (2.7); and define

$$f : \mathbf{R}^p \times \mathbf{R}^p \times \mathbf{R}^r \to \mathbf{R}^r$$

by $f(x,y,z) = f_0(x, f_0(y,z))$. Then f is trilinear and norm–preserving; and Hopf's construction

(2.20) $$\varphi : S^{2p-1} \to V_{2r,r}.$$

is a quadratic eigenmap.

These are not usually expressible in the form in (1.31):

(2.21) **Proposition.** *If $p \neq 1, 2, 4, 8$ then (2.20) admits no factorization of the form $\varphi = j \circ \psi$ where $\psi : S^{2p-1} \to S^{k-1}$ is an eigenmap and $j : S^{k-1} \to V_{2r,r}$ is a totally geodesic embedding.*

Proof. Suppose that such a factorization exists. Again we represent points $(x,y) \in S^{2p-1} \subset \mathbf{R}^p \times \mathbf{R}^p$; inspection of (2.17) shows that we can assume that the last component of ψ (as a map into $\mathbf{R}^k$) is $|x|^2 - |y|^2$; and $-j(x_0) = j(-x_0)$ for $x_0 = (0,\ldots,0,1) \in \mathbf{R}^k$. By Proposition (2.15), j is determined by an orthogonal multiplication $f : \mathbf{R}^k \times \mathbf{R}^r \to \mathbf{R}^{2r}$; and from (2.17), (2.19) we see that f has the form

$$f(w,z) = (f_1(w_1,\ldots,w_{k-1};z), w_k \cdot z),$$

where $w = (w_1,\ldots,w_k)$, $z \in \mathbf{R}^r$, and $f_1 : \mathbf{R}^{k-1} \times \mathbf{R}^r \to \mathbf{R}^r$ is an orthogonal multiplication. From Theorem (2.7) we find $k - 1 \leq \rho(r) = p$.

However, (2.17), (2.19) also show that ψ is Hopf's construction on an orthogonal multiplication $\tilde{f} : \mathbf{R}^p \times \mathbf{R}^p \to \mathbf{R}^{k-1}$. By Theorem (2.7), the existence of $\tilde{f}$ is compatible with $p \geq k - 1$ iff $p = 1,2,4,8$. //

3. POLYNOMIAL MAPS BETWEEN SPHERES

We present results of R. Wood [Wo2] – in particular, Theorems (3.4), (3.10), (3.19).

(3.1) We say that $\varphi : S^m \to S^n$ is a *polynomial map* if it is the restriction of a map $F : \mathbf{R}^{m+1} \to \mathbf{R}^{n+1}$ whose components $F^\alpha : \mathbf{R}^{m+1} \to \mathbf{R}$ are polynomials. φ (or F) is a *k–form* if each F^α is homogeneous of degree k.

(3.2) **Lemma.** *For n odd, the quadratic form $\theta : S^n \to S^n$ which is the restriction of*

$$F(x_1,\ldots,x_{n+1}) = (x_1^2 - x_2^2 - \ldots - x_{n+1}^2, 2x_1x_2, \ldots, 2x_1x_{n+1})$$

has Brouwer degree 2.

Proof. Write $z = x_1 + ix_2$, $w = (x_3 + ix_4, \ldots, x_n + ix_{n+1})$; in terms of these variables F becomes

$$F(z,w) = (z^2 - |w|^2, 2(Re\ z)w).$$

The homotopy

$$F_t(z,w) = (z^2 - (1-t)|w|^2, 2(1-t)(Re\ z)w + 2tiw)$$

carries $\mathbf{R}^{n+1} - \{0\}$ to itself for each $t (0 \leq t \leq 1)$; and deforms $F_0 = F$ to $F_1(z,w) = (z^2, 2iw)$. But F_1 is the $(n-1)$–fold suspension of the map $z \to z^2$, which has Brouwer degree 2. //

(3.3) Lemma. *Any two maps $\psi, \psi' : S^n \to S^n$ determine a third $\varphi : S^n \to S^n$ by $\varphi = -\psi + 2 < \psi, \psi' > \psi'$. If n is odd, then its homotopy class $[\varphi] = -[\psi] + 2[\psi']$.*

Proof. That $|\varphi(x)|^2 \equiv 1$ is immediate. To verify the second assertion, deform ψ and ψ' so that $\psi \equiv (0,\ldots,0,1)$ on the upper hemisphere S^n_+ and $\psi' \equiv (0,\ldots,0,1)$ on S^n_-. Then $\varphi|_{S^n_+}$ has the representation $(2\psi'^2_1 - 1, 2\psi'_1\psi'_2, \ldots, 2\psi'_1\psi'_{n+1})$; and because $2\psi'^2_1 - 1 = \psi'^2_1 - \psi'^2_2 - \ldots - \psi'^2_{n+1}$ on S^n, we conclude that $\varphi|_{S^n_+}$ is the composition of ψ' and the quadratic form θ of Lemma (3.2). Hence for n odd, it represents $2[\psi']$.

Similarly $\psi|_{S^n_-}$ has the representation $(\psi_1, -\psi_2, \ldots, -\psi_{n+1})$, which represents $-[\psi]$ for n odd. From the defintion of addition in $\pi_n(S^n)$ we conclude that $[\varphi] = -[\psi] + 2[\psi']$. //

(3.4) Theorem. *If n is odd, then every class $k \in \pi_n(S^n) = \mathbf{Z}$ can be represented by a polynomial $|k|$-form.*

Proof. If in Lemma (3.3) we have forms ψ, ψ' of polynomial degrees k, k', then $-|\psi'|^2\psi + 2 < \psi, \psi' > \psi'$ has degree $k + 2k'$. Since $\pm 1 \in \pi_n(S^n)$ can be represented by linear forms, the proof of (3.4) follows by induction, using Lemma (3.3). //

(3.5) If n is even we know of no class $\neq 0, \pm 1$ in $\pi_n(S^n)$ which has a polynomial representative. Therefore we prepare for non–existence results.

(3.6) Lemma. *If there are non–constant polynomial maps $\varphi : S^m \to S^r$ and $\psi : S^r \to S^n$, then there is a non–constant polynomial map $S^m \to S^n$.*

Proof. Provided that $\varepsilon > 0$ is sufficiently small, we can select points on S^r as follows:

1) $x, y \in \varphi(S^m)$ whose distance $d(x, y) = \varepsilon$;

2) $u, v \in S^r$ such that $d(u, v) = \varepsilon$ and $\psi(u) \neq \psi(v)$.

If $A \in O(r+1)$ is chosen to carry x, y to u, v respectively, then $\psi \circ A \circ \varphi : S^m \to S^n$ is a non–constant polynomial map. //

(3.7) Lemma. *If there is a non–constant polynomial map $\varphi : S^m \to S^n$, then there is a non–constant polynomial form $S^m \to S^n$.*

Proof. With θ as in Lemma (3.2), each component of the composition $\varphi \circ \theta$ is a sum of monomials of even degree. Multiplying each monomial by a suitable power of $|x|^2$ produces the required form. //

We need two results in number theory:

(3.8) Lemma [C]. *Over $\mathbb{R}$ it is not possible to represent*

$$y_1^2 + \ldots + y_{m+1}^2 \equiv f_1^2 + \ldots + f_m^2$$

with f_i rational functions of $y_1, \ldots, y_{m+1}$.

(3.9) Lemma [Pf]. *If m is a power of 2, then it is possible to represent*

$$(u_1^2 + \ldots + u_m^2)(v_1^2 + \ldots + v_m^2) = f_1^2 + \ldots + f_m^2,$$

with f_i are rational functions of $u_1, \ldots, u_m, v_1, \ldots, v_m$.

We are now in position to prove

(3.10) Theorem. *If m is a power of 2, then every polynomial map $S^m \to S^{m-1}$ is constant.*

Proof. By Lemma (3.7) it is sufficient to consider a k–form $F : \mathbb{R}^{m+1} \to \mathbb{R}^m$. Set $u = x_{m+1}, y = (x_1, \ldots, x_m)$; and write $F = R + uQ$, where R and Q involve only even powers of u.

Because $|F(x)|^2 \equiv |x|^{2k}$, we find

(3.11) $$|R|^2 + u^2|Q|^2 \equiv (u^2 + |y|^2)^k; \quad \text{and}$$

(3.12) $$<R, Q> \equiv 0.$$

If F were non–constant, we could assume $k \geq 1$ and that at least one component is not divisible by $|x|^2$. Replace u^2 by $-|y|^2$ in (3.11). Then $Q \not\equiv 0$; for otherwise Q would be divisible by $|x|^2$, and hence so would R.

Thus (3.11) can be written

(3.13) $$|y|^2 \equiv |R|^2/|Q|^2$$

where R, Q are now functions of y alone. By introducing a new variable y_{m+1}, we can use (3.12) to rewrite (3.13) in the form

(3.14) $$|y|^2 + y_{m+1}^2 \equiv |R + y_{m+1}Q|^2/|Q|^2.$$

Finally, set

(3.15) $$u = (R + y_{m+1}Q)/|Q|^2$$
$$v = Q.$$

If m is a power of 2, then from Lemma (3.9) we conclude that the right–hand member of (3.14) is expressible as $f_1^2 + \ldots f_m^2$, the sum of rational functions of the u_i, v_j.

However, (3.15) shows that the f_i are also rational functions of the y_j; thus (3.14) contradicts Lemma (3.8). //

(3.16) **Corollary.** *If $m \geq 2n$, then every polynomial map $\varphi : S^m \to S^n$ is constant.*

Proof. Observe that for every $n \geq 1$ there is an integer q such that $n < 2^q \leq 2n$. Now, suppose that there is a non–constant polynomial map $\varphi : S^m \to S^n, m \geq 2n$. Its restriction to a totally geodesic $S^{2^q} \subset S^m$ would be a non–constant polynomial map $\bar\varphi : S^{2^q} \to S^n$. Then composition $i \circ \bar\varphi$ of $\bar\varphi$ with the inclusion $i : S^n \to S^{2^q-1}$ would contradict Theorem (3.10). //

In a spirit similar to that of Theorem (3.10) we quote a result of Yiu:

(3.17) **Theorem [Y].** *If $n \geq 16$ is a power of 2, then any quadratic form $\varphi : S^{2n-1} \to S^n$ is constant. If $n = 2, 4, 8$, then any non–constant quadratic form $\varphi : S^{2n-1} \to S^n$ is (up to isometries) the Hopf fibration.*

(3.18) A quadratic form $\varphi : S^{p+q-1} \to S^n$ is a *Hopf form* if, modulo orthogonal transformations, it is obtained from the Hopf construction (2.3) on an orthogonal multiplication $f(p, q; n)$. We prove

(3.19) **Theorem.** *Every quadratic form $\varphi : S^m \to S^n$ is homotopic to a Hopf form.*

Proof. Let φ be the restriction of the form $F : \mathbb{R}^{m+1} \to \mathbb{R}^{n+1}$; thus

(3.20) $$|F(x)|^2 \equiv |x|^4.$$

By an orthogonal transformation we can assume that its first component

$$F^1(x) = x_1^2 + x_1 L + Q$$

where L and Q are linear and quadratic forms respectively in $x_2, \ldots, x_{m+1}$. Comparing the coefficients of x_1^3 in (3.20), we find $L \equiv 0$. We perform an orthogonal transformation on the space of $x_2, \ldots, x_{m+1}$ to diagonalize F^1:

$$F^1(x) = \sum_{i=1}^{m+1} \lambda_i x_i^2 \quad \text{with} \quad \lambda_1 = 1.$$

Now we partition the variables $x_1, \ldots, x_{m+1}$, putting those x_i with $\lambda_i = 1$ into one set and all others in the second. If the second is empty, then $F^1(x) = |x|^2$; from (3.20) we find $F^\alpha \equiv 0$ for $\alpha \geq 2$, so F is degenerate. We exclude that case.

Next, write

(3.21) $$F^1(x) = x_1^2 + \ldots + x_{k+1}^2 + \mu_1 y_1^2 + \ldots + \mu_q y_q^2$$

($k \geq 0, q \geq 1, \mu_i \neq 1$). Note that $-1 \leq \mu_i < 1$ by (3.20). We shall perform a homotopy carrying each μ_i into -1, and F into a Hopf form. For that we express $(F^2, \ldots, F^{n+1})$ as a vector function $A + B + 2C$, where

A is a quadratic form in $x = (x_1, \ldots, x_{k+1})$,

B is a quadratic form in $y = (y_1, \ldots, y_q)$; and

C is bilinear in x and y. Again comparing coefficients in (3.20) shows $A \equiv 0$; and

(3.22) $$|B|^2 + (\mu_1 y_1^2 + \ldots + \mu_q y_q^2)^2 \equiv |y|^4;$$

(3.23) $$<B, C> \equiv 0.$$

(3.24) $$2|C|^2 + |x|^2(\mu_1 y_1^2 + \ldots + \mu_q y_q^2) \equiv |x|^2 |y|^2.$$

Together with the restriction $\mu_i < 1$, those relations insure that for $0 \le t \le 1$

$$F_t = (F_t^1, (1-t)B + 2C)$$

is a homotopy in $\mathbf{R}^{n+1} - \{0\}$, where

$$F_t^1 = |x|^2 + (1-t)(\mu_1 y_1^2 + \ldots + \mu_q y_q^2) - t |y|^2.$$

It carries F to the quadratic form $(|x|^2 - |y|^2, 2C)$.

Finally, we write (3.24) as

$$2|C|^2 = |x|^2 \left[(1-\mu_1)y_1^2 + \ldots + (1-\mu_q)y_q^2\right].$$

Here we replace y_i by $y_i \sqrt{2}/\sqrt{2 - t(1+\mu_i)}$; that gives a homotopy $C_t (0 \le t \le 1)$ of $C = C_0$, where

$$|C_t|^2 \equiv |x|^2 \sum_{i=1}^{q} (1-\mu_i)y_i^2/(2 - t(1+\mu_i)).$$

But $|C_1|^2 \equiv |x|^2 |y|^2$, so the homotopy $(|x|^2 - |y|^2, 2C_t)$ in $\mathbf{R}^{n+1} - \{0\}$ carries our quadratic into a Hopf form, as claimed. //

4. Notes and comments

(4.1) We have followed [BGM] in presenting the basic properties of spherical harmonics leading to Theorem (1.9).

(4.2) The standard minimal immersions $\psi_{m,k}$ in (1.17) were constructed by Do Carmo and Wallach [dCW]. The proof via Proposition (1.15) is a slight modification of [dCW]; it was given by Takahashi [Ta] – and works when S^m is replaced by any compact irreducible G/H.

(4.3) The following rigidity theorem was proved by Calabi [Ca] for $m = 2$; and generalized to the form below in [dCW]; in the notation of (1.17),

Theorem. *Let $\psi : S^m(r) \to S^n$ be a full minimal Riemannian immersion. Then $r = c(k)$ and $n \le n(k)$. Moreover*

(a) *If $m = 2$ or $k \le 3$, then $\psi = \psi_{m,k}$, modulo orthogonal transformations.*

(b) *If $m \geq 3$ and $k \geq 4$, then the totality of equivalence classes of such immersions (under orthogonal transformations) is smoothly parametrized by a compact convex body L in a finite dimensional vector space W with $\dim W \geq 18$. The interior (resp., boundary) points of L correspond to immersions with $n = n(k)$ (resp., $n < n(k)$).*

(4.4) In a vein similar to (4.3) above, D'Ambra–Toth [TotA] have parametrized the equivalence classes of full eigenmaps by suitable compact convex sets L in finite dimensional vector spaces; and have given lower bounds on their dimensions. Toth has classified the full quadratic eigenmaps $\varphi : S^3 \to S^n$. All of these constructions are described in his monograph [Tot2].

(4.5) Real Clifford algebras are essentially the orthogonal multiplications $f(p, q; q)$. Relations between representations of Clifford algebras and orthogonal multiplications are discussed in detail in [ABS].

(4.6) An orthogonal multiplication $f(p, q; q)$ determines $(p-1)$ orthonormal vector fields on S^{q-1}. Indeed, we can normalize f so that $f(e_p, y) = y$ for all $y \in \mathbb{R}^q$, where $e_p = (0, \ldots, 0, 1) \in \mathbb{R}^p$; then the vector fields are $v_i(y) = f(e_i, y), 1 \leq i \leq p-1$. That relationship is due to Eckmann ([E], [GWY]). Adams [A2] proved that $\rho(q) - 1$ is the maximum number of linearly independent vector fields on S^{q-1} – with ρ as in (2.7). Note that that is zero when q is odd.

(4.7) Parker [Pa] has classified the $f(p, p; r)$ for $p = 2, 3$. Toth [Tot2] has parametrized the range–equivalence classes of full $f(p, p; r)$ by a compact convex body in a finite dimensional vector space, as in (4.3), (4.4).

(4.8) Proposition (2.15) appears in [Sm1].

(4.9) The Hopf construction $\varphi : S^{2p-1} \to S^n$ on an $f(p, p; n)$ is a harmonic morphism iff $p = n = 1, 2, 4, 8$ [Gi].

(4.10) Yiu [Y] has shown that every quadratic form $S^{25} \to S^{23}$ is constant.

(4.11) Wang [Wa] has observed that every homotopy class in $\pi_{p-1}(O(\infty))$ has a totally geodesic embedded representative of constant curvature, as in (2.2). Here $O(\infty) = \lim\limits_{n \to +\infty} O(n)$, induced from the standard inclusion $O(n) \hookrightarrow O(n+1)$.

(4.12) Loday [Lo] has shown that *every polynomial map $S^1 \times \ldots \times S^1 = T^n \to S^n$ is contractible if $n \geq 2$.* That has been generalized (with a different proof) by Bochnak–Kucharz [BK], as follows: *Let $n_1, \ldots, n_k$ be positive integers ($k \geq 2$); set $n = n_1 + \ldots + n_k$. Then every polynomial map $S^{n_1} \times \ldots \times S^{n_k} \to S^n$ is contractible iff at least two of the n_i are odd.*

(4.13) With the same notation, *let $F : \mathbb{R}^{n_1+1} \times \ldots \times \mathbb{R}^{n_k+1} \to \mathbb{R}^{n+1}$ be a polynomial map sending $S^{n_1} \times \ldots \times S^{n_k} \to S^n$. If F is homogeneous in each variable, then its restriction*
$S^{n_1} \times \ldots \times S^{n_k} \to S^n$ *is contractible* [BK].

CHAPTER IX. EXISTENCE OF HARMONIC JOINS

INTRODUCTION

Let $u : S^{p-1} \to S^{q-1}, v : S^{r-1} \to S^{s-1}$ be two maps of spheres; their *join* is a map $u * v : S^{p+r-1} \to S^{q+s-1}$. If u, v are eigenmaps, then the condition of harmonicity for $u * v$ reduces to an O.D.E.; more precisely, to a pendulum type equation with variable gravity and damping, with specified asymptotic limits.

In Section 1 we derive the above mentioned reduction equation, following the theoretical lines set out in Chapter IV. Next, we establish conditions (the *damping conditions*) which are necessary and sufficient for the existence of equivariant harmonic joins. Amongst the applications, we note the following: *Each element of the group $\pi_n(S^n) = \mathbb{Z}$ can be represented by a harmonic map between Euclidean spheres, provided $n \leq 7$ or $n = 9$*. These results can be obtained by a direct qualitative study of the pendulum equation, based on comparison arguments. That was the approach in the basic work of Smith [Sm1,2], where these reduction techniques were first applied and existence obtained in the most important cases. Here we follow a method of Ding [Di], which combines the above qualitative study with the variational theory of the reduced energy functional J. This leads us naturally to consider certain ellipsoidal deformations of S^n; in particular, we show that *each element of $\pi_n(S^n) = \mathbb{Z}$ can be represented harmonically provided that S^n is given a suitable ellipsoidal metric, for every $n \geq 1$* (see (5.3), (6.2) below for the explicit characterizations of these metrics).

1. THE REDUCTION EQUATION

(1.1) For $a, b > 0$ we introduce the ellipsoid

$$Q^{p+r-1}(a,b) = \{(x,y) \in \mathbb{R}^p \times \mathbb{R}^r : |x|^2/a^2 + |y|^2/b^2 = 1\} .$$

We call b/a its *dilatation*. It is convenient to parametrize the points of $Q^{p+r-1}(a,b)$ by

(1.2) $\qquad (a \sin s \cdot x, b \cos s \cdot y)$

with $x \in S^{p-1}, y \in S^{r-1}$ and $0 \leq s \leq \pi/2$.

Clearly, $Q^{p+r-1}(1,1) = S^{p+r-1} = S^{p-1} * S^{r-1}$.

The Riemannian metric on $Q^{p+r-1}(a,b)$ induced from its embedding in $\mathbb{R}^{p+r}$ is

(1.3) $\qquad g = a^2 \sin^2 s \cdot g_{p-1} + b^2 \cos^2 s \cdot g_{r-1} + h^2(s)ds^2$

where $h^2(s) = a^2 \cos^2 s + b^2 \sin^2 s$; and g_m denotes the standard metric of S^m.

(1.4) Let $\alpha : [0, \pi/2] \to [0, \pi/2]$ be a smooth function satisfying the boundary conditions

(1.5) $$\alpha(0) = 0, \quad \alpha(\pi/2) = \pi/2 .$$

For given $a, b, c, d > 0$ the *equivariant ellipsoidal α–join* of two eigenmaps $u : S^{p-1} \to S^{q-1}$ and $v : S^{r-1} \to S^{s-1}$ with eigenvalues λ_u and λ_v is the map
$$u * v = u *_\alpha v : Q^{p+r-1}(a, b) \to Q^{q+s-1}(c, d)$$
given by

(1.6) $$(a \sin s \cdot x, b \cos s \cdot y) \to (c \sin \alpha(s) \cdot u(x), d \cos \alpha(s) \cdot v(y)) .$$

(1.7) We have Riemannian submersions
$$\rho : Q^{p+r-1}(a, b) \to ([0, \pi/2], h^2(s)ds^2) = P$$
with $\rho(x, y, s) = s$; and similarly
$$\sigma : Q^{q+s-1}(c, d) \to ([0, \pi/2], k^2(s)ds^2) = Q ,$$
where as in (1.3)

(1.8) $$k^2(s) = c^2 \cos^2 s + d^2 \sin^2 s .$$

The diagram
$$\begin{array}{ccc} Q^{p+r-1}(a,b) & \xrightarrow{u*v} & Q^{q+s-1}(c,d) \\ \rho \downarrow & & \sigma \downarrow \\ P & \xrightarrow{\alpha} & Q \end{array}$$

is commutative. The fibres of ρ (resp., σ) over interior points are homogeneous, being orbits of the standard isometric action of $SO(p) \times SO(r)$ on $Q^{p+r-1}(a, b)$ (resp., $SO(q) \times SO(s)$ on $Q^{q+s-1}(c, d)$. Therefore, both ρ and σ have basic tension fields.

It is clear that $u * v$ is horizontal and satisfies (ii) in IV (4.13). Therefore, according to IV (4.14), $u * v$ is harmonic iff

(1.9) $$\ddot\alpha(s) + D(s)\dot\alpha(s) - G(s, \alpha, \dot\alpha) = 0$$

where $D(\rho) = \tau(\rho)$ and $G = \text{Trace}\ (u * v)^*(\nabla d\sigma)$.

Computations show that

(1.10)
$$D = (p-1)\cot - (r-1)\tan - \frac{h'}{h}$$
$$G = h^2\left[\frac{c^2\lambda_u}{a^2\sin^2} - \frac{d^2\lambda_v}{b^2\cos^2}\right]\frac{\sin\alpha\cos\alpha}{k^2(\alpha)} - \frac{k'(\alpha)}{k(\alpha)}\dot\alpha^2;$$

here and in the future we may abbreviate sin s by sin, $\alpha(s)$ by α, etc.

Henceforth we assume $a = 1 = b$, for simplicity. Substituting (1.10) with $h = 1$ in (1.9) gives

(1.11) **Reduction theorem.** *The ellipsoidal α-join*

$$u *_\alpha v : S^{p+r-1} \to Q^{q+s-1}(c,d)$$

is harmonic iff

$$\ddot\alpha + [(p-1)\cot - (r-1)\tan]\dot\alpha + \left[\frac{d^2\lambda_v}{\cos^2} - \frac{c^2\lambda_u}{\sin^2}\right]\frac{\sin\alpha\cos\alpha}{k^2(\alpha)} + \frac{k'(\alpha)}{k(\alpha)}\dot\alpha^2 = 0,$$

where α satisfies the boundary conditions (1.5).

(1.12) An elementary calculation shows that the harmonicity equation in Theorem (1.11) is the Euler–Lagrange equation of the 1–dimensional functional

(1.13) $$J(\alpha) = \int_0^{\pi/2}\left[k^2(\alpha)\dot\alpha^2 + \frac{c^2\sin^2\alpha}{\sin^2}\lambda_u + \frac{d^2\cos^2\alpha}{\cos^2}\lambda_v\right]\nu\,ds,$$

where $\nu = \sin^{p-1}\cos^{r-1}$. Up to a constant factor, $J(\alpha)$ coincides with the energy $E(u *_\alpha v)$.

(1.14) An equivalent expression of equation (1.11) is the divergence form

$$\frac{d}{ds}(k(\alpha)\dot\alpha\nu) = \left(\frac{c^2\lambda_u}{\sin^2} - \frac{d^2\lambda_v}{\cos^2}\right)\left(\frac{\sin\alpha\cos\alpha}{k(\alpha)}\right)\nu.$$

2. PROPERTIES OF THE REDUCED ENERGY FUNCTIONAL J

(2.1) Define the Hilbert space

$$A = \left\{ \alpha \in \mathcal{L}_1^2([0, \pi/2], \mathbb{R}) : \|\alpha\|^2 = \int_0^{\pi/2} (\dot{\alpha}^2 + \alpha^2)\nu \, ds < \infty \right\}.$$

For $p, r > 2$ the functional $J : \mathcal{A} \to \mathbb{R}$ is defined and smooth on $\mathcal{A}$. That is a consequence of the following Sobolev inequalities for the Riemannian manifolds

$$\left([0, \pi/2], \sin^{p-3} \cos^{r-1}\right), \left([0, \pi/2], \sin^{p-1} \cos^{r-3}\right):$$

For a suitable constant $C > 0$,

(2.2) $$\left. \begin{array}{l} \int_0^{\pi/2} \alpha^2 \sin^{p-3} \cos^{r-1} ds \\ \int_0^{\pi/2} \alpha^2 \sin^{p-1} \cos^{r-3} ds \end{array} \right\} \leq C \int_0^{\pi/2} (\dot{\alpha}^2 + \alpha^2) \sin^{p-1} \cos^{r-1} ds.$$

If either $p = 2$ or $r = 2$, we extend the definition of J, allowing it to assume the value $+\infty$. However, most of the subsequent analysis will be carried out in case $p, r > 2$; and extended to either case $p = 2$ or $r = 2$ by continuity arguments.

(2.3) The directional derivative of J at α in the direction $\xi \in \mathcal{A}$ is

$$dJ(\alpha)\xi = 2 \int_0^{\pi/2} \left\{ k^2(\alpha) \dot{\alpha} \, \dot{\xi} + \left[k(\alpha) k'(\alpha) \dot{\alpha}^2 \right. \right.$$
$$\left. \left. + \left(\frac{c^2 \lambda_u}{\sin^2} - \frac{d^2 \lambda_v}{\cos^2} \right) \sin \alpha \cos \alpha \right] \xi \right\} \nu \, ds.$$

In order to investigate the critical points of J subject to the boundary conditions (1.5), it is convenient to introduce the *closed, convex subset*

(2.4) $\quad A_0 = \{\alpha \in \mathcal{A} : 0 \leq \alpha(s) \leq \pi/2 \text{ for } 0 \leq s \leq \pi/2\}$.

If $\underline{\alpha} \in A_0$ is a critical point of $J|_{A_0}$, then it is not difficult to show that $\underline{\alpha}$ satisfies the *Euler–Lagrange equation* (1.11) characterizing harmonicity. Observe that $\alpha_0 \equiv 0$ and $\alpha_{\pi/2} \equiv \pi/2$ are both critical points of J.

The following two are standard properties of integrals $I : \mathcal{L}_1^2(M, N) \to \mathbb{R}$ of the form

$$I(\varphi) = \int_M \left[A(x, \varphi(x)) |d\varphi(x)|^2 + B(x, \varphi(x)) \right] dx,$$

where M, N are compact, $A, B : M \times N \to \mathbb{R}$ are smooth functions, and $A > 0$.

(2.5) Lemma. *For $p, r > 2$ the functional $J : \mathcal{A} \to \mathbb{R}$ is weakly lower semi-continuous.*

I.e., for any sequence $\underline{\alpha}, (\alpha_i)_{i \geq 1}$ in $\mathcal{A}$ such that $< \alpha_i, \beta > \to < \underline{\alpha}, \beta >$ for all $\beta \in \mathcal{A}$, then
$$J(\underline{\alpha}) \leq \liminf_{i \to \infty} J(\alpha_i) .$$

(2.6) Corollary. *J assumes its minimum $\underline{\alpha}$ on $\mathcal{A}_0$:*
$$J(\underline{\alpha}) = \inf \{ J(\alpha) : \alpha \in \mathcal{A}_0 \} .$$

Proof. Take a minimizing sequence $(\alpha_i)_{i \geq 1}$ in $\mathcal{A}_0$. Because $J(\alpha_i)$ is bounded and each $\alpha_i \in \mathcal{A}_0$, $(\|\alpha_i\|)_{i \geq 1}$ is bounded. Thus a subsequence of $(\alpha_i)_{i \geq 1}$ subconverges weakly and we apply (2.5). //

(2.7) Lemma (Palais–Smale condition). *Assume that $p, r > 2$. If $(\alpha_i)_{i \geq 1} \subset \mathcal{A}_0$ is a sequence on which J is bounded and its differential $dJ(\alpha_i) \to 0$ as $i \to \infty$, then a subsequence of $(\alpha_i)_{i \geq 1}$ converges in $\mathcal{A}_0$.*

Proof. Step 1. First we assume $c = 1 = d$; thus $k^2(\alpha) \equiv 1$. We have noted in (2.1) that J is smooth on $\mathcal{A}$. Now we observe that $(\|\alpha_i\|)$ is bounded, because $(J(\alpha_i))$ is and each $\alpha_i \in \mathcal{A}_0$. Therefore a subsequence, still called (α_i), converges weakly to some $\tilde{\alpha} \in \mathcal{A}_0$. That insures

$$(2.8) \qquad \int_0^{\pi/2} (\alpha_i - \alpha_j)^2 \nu \, ds \to 0 \quad \text{as} \quad i, j \to \infty .$$

From (2.3) we see that

$$dJ(\alpha_i)(\alpha_i - \alpha_j) = 2 \int_0^{\pi/2} \left[\dot{\alpha}_i (\dot{\alpha}_i - \dot{\alpha}_j) + L \sin \alpha_i \cos \alpha_i (\alpha_i - \alpha_j) \right] \nu \, ds ,$$

where
$$L = \frac{\lambda_u}{\sin^2} - \frac{\lambda_v}{\cos^2} .$$

Expressing $dJ(\alpha_j)(\alpha_i - \alpha_j)$ similarly, taking their difference and using the hypothesis that those directional derivatives are $o(1)$ (i.e., they tend to 0 as $i, j \to \infty$), we find

(2.9) $$\begin{aligned}o(1) &= (dJ(\alpha_i) - dJ(\alpha_j))(\alpha_i - \alpha_j) \\ &= \int_0^{\pi/2} (\dot\alpha_i - \dot\alpha_j)^2 \nu \, ds \\ &\quad + \int_0^{\pi/2} [L(\sin\alpha_i \cos\alpha_i - \sin\alpha_j \cos\alpha_j)(\alpha_i - \alpha_j)] \nu \, ds .\end{aligned}$$

The second integral is $o(1)$; that is seen by writing it as a sum over $[0,\varepsilon]$, $[\pi/2 - \varepsilon, \pi/2]$, and $[\varepsilon, \pi/2 - \varepsilon]$; and estimating each separately. In fact, the absolute value of the second integral is dominated by

$$\int_0^{\pi/2} |L|(\alpha_i - \alpha_j)^2 \nu \, ds = \int_0^\varepsilon + \int_{\pi/2-\varepsilon}^{\pi/2} + \int_\varepsilon^{\pi/2-\varepsilon}$$
$$\leq \int_0^\varepsilon + \int_{\pi/2-\varepsilon}^{\pi/2} |L| \frac{\pi^2}{4} \nu \, ds + c(\varepsilon) \int_0^{\pi/2} (\alpha_i - \alpha_j)^2 \nu \, ds .$$

Because $p, r > 2$, the contribution of

$$\int_0^\varepsilon \quad \text{and} \quad \int_{\pi/2-\varepsilon}^{\pi/2}$$

can be made arbitrarily small by taking ε small.

On the other hand, by fixing ε and taking i, j sufficiently large, the remaining integral can be made arbitrarily small, by (2.8).

Using that in (2.9) gives

(2.10) $$o(1) = \int_0^{\pi/2} (\dot\alpha_i - \dot\alpha_j)^2 \nu \, ds .$$

But (2.8) and (2.10) together show that (α_i) converges strongly in $\mathcal{A}_0$; that ends the proof in case $c = 1 = d$.

Step 2. If $d/c \neq 1$, it is convenient to change coordinates on $Q^{q+s-1}(c,d)$. Let

$$t = P(s) = \int_0^s k(\sigma)d\sigma \qquad (0 \leq s \leq \pi/2) .$$

In terms of the join coordinates (w, z, t) on $S^{q-1} * S^{s-1}$ with $0 \leq t \leq P(\pi/2)$, the metric g on $Q^{q+s-1}(c,d)$ is expressed by

$$g = c^2 f_1^2(t) g_{q-1} + d^2 f_2^2(t) g_{s-1} + dt^2 ,$$

where
$$f_1(t) = \sin P^{-1}(t), \quad f_2(t) = \cos P^{-1}(t).$$

Then the energy functional takes the form
$$\hat{J}(\beta) = \int_0^{\pi/2} \left[\dot\beta^2 + \frac{c^2 f_1^2(\beta)}{\sin^2} \lambda_u + \frac{d^2 f_2^2(\beta)}{\cos^2} \lambda_v \right] \nu \, ds \, ;$$

and by construction
$$J(\alpha) = \hat{J}(P(\alpha)).$$

Because f_1 and f_2 behave qualitatively like sin and cos, the Palais–Smale condition for $\hat{J}$ can be verified as in Step 1. //

The following is a consequence of the Morse theory of critical points of smooth functions adjusted to include domains which are closed convex subsets of Banach spaces [CE], [St]:

(2.11) **Lemma** (*Mountain pass*). *Assume that* $p, r > 2$. *If* α_0 *is an isolated local minimum for* $J|_{\mathcal{A}_0}$ *and there is an* $\alpha \neq \alpha_0$ *in* $\mathcal{A}_0$ *such that* $J(\alpha) = J(\alpha_0)$, *then there is a critical point* $\beta \in \mathcal{A}_0$ *with* $J(\beta) > J(\alpha_0)$. *Moreover, if* $J|_{\mathcal{A}_0}$ *has two isolated local minima, then it has another critical point (which is not an absolute minimum).*

(2.12) For any critical point α and variation ξ, the corresponding *second variation of J* is

$$\frac{1}{2} \nabla^2 J(\alpha)(\xi,\xi) = \int_0^{\pi/2} \left\{ \left[(k'(\alpha)^2 + k(\alpha)k''(\alpha))\dot\alpha^2 + \left(\frac{c^2 \lambda_u}{\sin^2} - \frac{d^2 \lambda_v}{\cos^2} \right) \cos 2\alpha \right] \xi^2 + k(\alpha) \left[4k'(\alpha)\dot\alpha \xi \dot\xi + k(\alpha)\dot\xi^2 \right] \right\} \nu \, ds.$$

We say that α is *stable* if $\nabla^2 J(\alpha)(\xi,\xi) > 0$ for every variation ξ.

3. ANALYSIS OF THE O.D.E.

(3.1) **Proposition.** *Let α be a non–constant critical point of $J|_{\mathcal{A}_0}$. Then $\dot\alpha(s) > 0$ for every $0 < s < \pi/2$, and α satisfies the boundary conditions (1.5). Otherwise said, α provides an equivariant harmonic join.*

Proof. This requires a detailed study of the qualitative properties of solutions of the pendulum equation (1.11).

Step 1. We consider that equation in the more general context of maps $\varphi : Q^{p+r-1}(a,b) \to Q^{q+s-1}(c,d)$:

(3.2)
$$\ddot{\alpha} + \left[(p-1)\cot - (r-1)\tan - \frac{h}{h}\right]\dot{\alpha} + \frac{h^2}{k^2(\alpha)}\left(\frac{d^2\lambda_v}{b^2\cos^2} - \frac{c^2\lambda_u}{a^2\sin^2}\right)\sin\alpha\cos\alpha + \frac{k'(\alpha)}{k(\alpha)}\dot{\alpha}^2 = 0,$$

where h is as in (1.3); in particular, h is an analytic function which is bounded and bounded away from 0. (Note that (3.2) is (1.9).)

We apply the transformation $\tan s = e^t (t \in \mathbb{R})$. With the notation

$$A(t) = \alpha(\tan^{-1} e^t) \quad \text{and} \quad H(t) = h(\tan^{-1} e^t),$$

equation (3.2) becomes

(3.3)
$$A'' + \left[\frac{(p-2)e^{-t} - (r-2)e^t}{e^t + e^{-t}} - \frac{H'}{H}\right]A' =$$
$$\left[(c^2 - d^2)A'^2 + \frac{H^2}{e^t + e^{-t}}\left(\frac{c^2\lambda_u e^{-t}}{a^2} - \frac{d^2\lambda_v e^t}{b^2}\right)\right]\frac{\sin A \cos A}{k^2(A)}$$

with boundary conditions for $A : \mathbb{R} \to [0, \pi/2]$

(3.4)
$$\lim_{t \to -\infty} A(t) = 0, \quad \lim_{t \to +\infty} A(t) = \pi/2.$$

Step 2. $0 < A(t) < \pi/2$ for all $t \in \mathbb{R}$.

For if $A(\bar{t}) = 0$ for some $\bar{t} \in \mathbb{R}$, then $A'(\bar{t}) \neq 0$; for otherwise $A \equiv 0$. Thus A would assume negative values, and consequently α could not belong to $\mathcal{A}_0$. Similarly, A does not assume the value $\pi/2$.

Step 3. $A' > 0$ on $\mathbb{R}$: Let t_0 be the solution of $c^2\lambda_u e^{-t}/a^2 = d^2\lambda_v e^t/b^2$; and suppose $A'(\bar{t}) = 0$ for some $\bar{t} \leq t_0$. Because A is analytic and non-constant, the zeroes of A' are isolated, so there is an $\varepsilon > 0$ such that $A'(t) \neq 0$ for $\bar{t} - \varepsilon < t < \bar{t}$.

Consider the linear equation

(3.5)
$$Y'(t) + P_A(t)Y(t) = Q_A(t),$$

where

$$P_A(t) = 2\left[\frac{A''}{A'} + \frac{(p-2)e^{-t} - (r-2)e^t}{e^t + e^{-t}} - \frac{H'}{H}\right] + \frac{(d^2 - c^2)\sin A \cos A}{k^2(A)}A'$$

$$Q_A(t) = 2H^2\left[\frac{c^2\lambda_u e^{-t}}{a^2} - \frac{d^2\lambda_v e^t}{b^2}\right]\frac{\sin A \cos A}{A'(e^t + e^{-t})k^2(A)}.$$

Then $P_A(t) \equiv Q_A(t)$ on $(\bar{t} - \varepsilon, \bar{t})$ because A is a solution of (3.3). Therefore $\overline{Y}(t) \equiv 1$ is a solution of (3.5) on $(\bar{t} - \varepsilon, \bar{t})$; and so is expressible as

$$\overline{Y}(t) \equiv 1 \equiv \frac{\int_{\tilde{t}}^{t} Q_A(\rho) \exp\left(\int_{\tilde{t}}^{\rho} P_A(w)dw\right) d\rho + C}{\exp\left(\int_{\tilde{t}}^{t} P_A(w)dw\right)} \tag{3.6}$$

for some $\bar{t} - \varepsilon < \tilde{t} < \bar{t}$ and $C \in \mathbb{R}$.

If T is the first point where $-\infty \leq T < \bar{t}$ and $A'(T) = 0$, then (3.6) holds for $T < t < \bar{t}$. Performing the integrations in (3.6) we find

$$1 \equiv N(t)/D(t) \tag{3.7}$$

where

$$N(t) = \int_{\tilde{t}}^{t} \left(\frac{c^2 \lambda_u e^{-\rho}}{a^2} - \frac{d^2 \lambda_v e^{\rho}}{b^2} \right) \frac{(1+e^{-2\rho})^{2-p}(1+e^{2\rho})^{2-r} \sin 2A}{e^{\rho} + e^{-\rho}} A' d\rho + C$$

and

$$D(t) = A'^2 (1+e^{-2t})^{2-p}(1+e^{2t})^{2-r} \frac{k^2(A)}{H^2(t)}.$$

Thus for $T < t < \bar{t}$ we have

$$N(t) > 0. \tag{3.8}$$

$$N'(t) \neq 0 \quad \text{because} \quad A' \neq 0, \ 0 < A < \pi/2 \quad \text{and} \quad \bar{t} \leq t_0. \tag{3.9}$$

Furthermore

$$T = -\infty. \tag{3.10}$$

For otherwise $D(T) = 0$, and so $N(T) = 0$, by (3.7). Using $N(\bar{t}) = 0$ and (3.8), we see that N must have an interior maximum on $[T, \bar{t}]$, contradicting (3.9).

We conclude from (3.10) that $A' \neq 0$ on $(-\infty, \bar{t})$ and that (3.7) holds there. But there must be points $\tilde{\tilde{t}} \in (-\infty, \bar{t})$ at which $A'(\tilde{\tilde{t}})$ is arbitrarily close to 0; for if A' is bounded away from 0, the values of the solution A would not remain in $[0, \pi/2]$. Thus D, and consequently N, must have values arbitrarily close to 0. Together with $N(\bar{t}) = 0$ and (3.8), that insures that N has a local maximum in $(-\infty, \bar{t})$, contradicting (3.9). Similarly, there is no $\bar{t} > t_0$ for which $A'(\bar{t}) = 0$. So $A' \neq 0$ on

R, and (3.7) holds for every $t \in \mathbf{R}$. And $A' < 0$ on **R** is not possible. For otherwise t_0 would be a minimum of N, again leading to a contradiction.

Step 4. Because $A' > 0$ on **R**, $\lim_{t \to \pm\infty} A(t)$ exist.

The inequalities $0 < A(t) < \pi/2$ insure that, for any small $\varepsilon > 0$ and large $C > 0$, there exists $\tilde{t} > C$ (or $\tilde{t} < -C$) with

$$A'(\tilde{t}) < \varepsilon \quad \text{and} \quad |A''(\tilde{t})| < \varepsilon .$$

Otherwise, A would go out of bounds. Simple inspection of (3.3) now shows that the only limits possible are those in (3.4). //

(3.11) **Lemma** (*a priori estimates*). *If $\alpha \in \mathcal{A}_0$ is a non-constant critical point, then*

$$J(\alpha) < J(\alpha_0) .$$

Proof.

$$J(\alpha) - J(\alpha_0) = \int_0^{\pi/2} \left[k^2(\alpha) \dot{\alpha}^2 + \left(\frac{c^2 \lambda_u}{\sin^2} - \frac{d^2 \lambda_v}{\cos^2} \right) \sin^2 \alpha \right] \nu \, ds .$$

But from (1.14)

$$\left(\frac{c^2 \lambda_u}{\sin^2} - \frac{d^2 \lambda_v}{\cos^2} \right) (\sin^2 \alpha) \nu = k(\alpha) \tan \alpha \frac{d}{ds} (k(\alpha) \dot{\alpha} \, \nu)$$

$$= \frac{d}{ds} \left[k^2(\alpha)(\tan \alpha) \dot{\alpha} \, \nu \right] - k^2(\alpha) \left(\frac{k'(\alpha)}{k(\alpha)} \tan \alpha + \frac{1}{\cos^2 \alpha} \right) \dot{\alpha}^2 \nu ,$$

so

$$J(\alpha) - J(\alpha_0) = \int_0^{\pi/2} \left[1 - \frac{k'(\alpha)}{k(\alpha)} \tan \alpha - \frac{1}{\cos^2 \alpha} \right] k^2(\alpha) \dot{\alpha}^2 \nu \, ds$$
$$+ k^2(\alpha)(\tan \alpha) \dot{\alpha} \nu \big|_0^{\pi/2} .$$

This last term vanishes: indeed, the asymptotic behaviour of α is qualitatively the same as in the case $k^2(\alpha) \equiv 1$; therefore we can use the estimates established in X (1.25) below; these yield the desired vanishing. Now, an elementary computation starting with $k^2(\alpha) = d^2 \sin^2 \alpha + c^2 \cos^2 \alpha$ shows that

$$k^2(\alpha) \left[1 - \frac{k'(\alpha)}{k(\alpha)} \tan \alpha - \frac{1}{\cos^2 \alpha} \right] = -d^2 \tan^2 \alpha ,$$

so

$$J(\alpha) - J(\alpha_0) = -d^2 \int_0^{\pi/2} \tan^2 \alpha \, \dot{\alpha}^2 \, \nu \, ds < 0 . //$$

In the terminology of (2.12) and (3.1) we have

(3.12) **Proposition.** *Suppose $J(\alpha_{\pi/2}) \geq J(\alpha_0)$. Then there is an equivariant harmonic join iff α_0 is an unstable critical point of J.*

Proof. Let $\underline{\alpha} \in \mathcal{A}_0$ be a minimum of J, as in Corollary (2.6). If α_0 is unstable, then clearly $\underline{\alpha} \neq \alpha_{\pi/2}, \alpha_0$. Then $\underline{\alpha}$ provides a harmonic join, by Proposition (3.1).

Conversely, assume first $p, r > 2$; and suppose that α_0 is stable. Let $\tilde{\alpha}$ provide an equivariant harmonic join. By Lemma (3.11) we have $J(\tilde{\alpha}) < J(\alpha_0)$; but application of the mountain pass Lemma (2.11) to the closed convex set

$$\tilde{\mathcal{A}}_0 = \{\alpha \in \mathcal{A}_0 : 0 \leq \alpha(s) \leq \tilde{\alpha}(s) \quad \text{for all} \quad 0 \leq s \leq \pi/2\}$$

shows the existence of a critical point $\beta \in \tilde{\mathcal{A}}_0$ with $J(\beta) > J(\alpha_0)$. But, again by Proposition (3.1), β provides an equivariant harmonic join, contradicting Lemma (3.11).

Finally, for a minimum $\underline{\alpha}$ of J, the condition $J(\underline{\alpha}) < J(\alpha_0)$ is *open*; i.e., it is satisfied for nearby values of the parameters $p, r, \lambda_u, \lambda_v, c, d$. We apply that to verify Proposition (3.12) in the cases $p = 2, r > 2$ and $p = 2 = r$. //

Note that the condition $J(\alpha_{\pi/2}) \geq J(\alpha_0)$ can be achieved by interchanging the roles of the eigenmaps u, v if necessary.

4. THE DAMPING CONDITIONS

The main goal of this Section is to prove

(4.1) **Theorem.** *Assume that $p, r \geq 2$. Let $u : S^{p-1} \to S^{q-1}$ and $v : S^{r-1} \to S^{s-1}$ be eigenmaps with eigenvalues λ_u and λ_v. Then there exists an equivariant harmonic join*

$$u *_\alpha v : S^{p+r-1} \to Q^{q+s-1}(c, d)$$

iff the following damping conditions hold:

$(DC)_1 \quad \begin{cases} \text{(i)} \quad (r-2)^2 < 4\lambda_v d^2/c^2 \\ \\ or \\ \\ \text{(ii)} \quad \sqrt{(p-2)^2 + 4\lambda_u} + \sqrt{(r-2)^2 - 4\lambda_v d^2/c^2} < p + r - 4; \end{cases}$

and

$$(DC)_2 \begin{cases} \text{(i)} & (p-2)^2 < 4\lambda_u c^2/d^2 \\ \text{or} \\ \text{(ii)} & \sqrt{(r-2)^2 + 4\lambda_v} + \sqrt{(p-2)^2 - 4\lambda_u c^2/d^2} < p + r - 4 . \end{cases}$$

We prepare the way to its proof: firstly, the following assertion can be obtained by integrating by parts:

(4.2) **Lemma.** $J(\alpha_{\pi/2}) \geq J(\alpha_0)$ iff

(4.3) $\quad \int_0^{\pi/2} \left(\dfrac{c^2 \lambda_u}{\sin^2} - \dfrac{d^2 \lambda_v}{\cos^2} \right) \nu \, ds \geq 0 \quad \text{iff} \quad (r-2)\lambda_u \geq (p-2)\lambda_v d^2/c^2 .$

(4.4) **Lemma.** *The constant solution α_0 is unstable if $(r-2)^2 < 4\lambda_v d^2/c^2$.*

Proof. For a variation $\xi \in \mathcal{A}$ with $\xi \geq 0$, we write $H(\xi) = \nabla^2 J(\alpha_0)(\xi, \xi)$. From (2.12) we obtain (up to a constant factor)

$$H(\xi) = \int_0^{\pi/2} \left[\dot{\xi}^2 + \left(\dfrac{\lambda_u}{\sin^2} - \dfrac{d^2 \lambda_v}{c^2 \cos^2} \right) \xi^2 \right] \nu \, ds .$$

First we suppose $r > 2$ and take $\xi = \sin \cos^{-q}$, where the real number $q \in (0, (r-2)/2)$ is to be determined. Then

(4.5) $\quad \begin{aligned} H(\xi) = & \int_0^{\pi/2} \left(q^2 \sin^2 - \dfrac{d^2 \lambda_v}{c^2} \right) \sin^{p+1} \cos^{r-2q-3} ds + \\ & \int_0^{\pi/2} \left[\left(\dfrac{\cos^2}{\sin^2} + 2q \right) + \dfrac{\lambda_u}{\sin^2} \right] \sin^{p+1} \cos^{r-2q-1} ds . \end{aligned}$

As a function of q, the second integral in (4.5) remains bounded as $q \to (r-2)/2$.

We show that the first integral tends to $-\infty$ as $q \to (r-2)/2$; thus for q close to $(r-2)/2$ we have $H(\xi) < 0$, so α_0 is unstable. Indeed, the first integral is majorized by

(4.6) $\quad \int_0^{\pi/2} \left[\dfrac{(r-2)^2}{4} - \dfrac{d^2 \lambda_v}{c^2} \right] \sin^{p+1} \cos^{r-2q-3} ds .$

The term in the brackets is negative. The exponent of cos tends to -1 as $q \to (r-2)/2$; and the integral $\int_{\pi/2-\epsilon}^{\pi/2} \cos^{-1} ds$ is divergent.

If $r = 2$, then $J(\alpha_0) = +\infty$, so α_0 is unstable in this case as well. //

(4.7) **Remark.** In the proof we used *unbounded* variations; i.e., ones with $\lim_{s \to \pi/2} \xi(s) = +\infty$.

Thus, even for small t, $\alpha_0 + t\xi \notin \mathcal{A}_0$. Formally, that means that we have shown that α_0 is unstable for $J : \mathcal{A} \to \mathbb{R}$, whereas we want to establish that α_0 is unstable for $J|_{\mathcal{A}_0}$. That can be done by defining for $M > 0$,

$$\xi_M(s) = \begin{cases} \xi(s) & \text{if } \xi(s) \le M \\ M & \text{if } \xi(s) > M. \end{cases}$$

Then ξ_M is an admissible variation for $J|_{\mathcal{A}_0}$; and $\lim_{M \to \infty} H(\xi_M) = H(\xi)$.

In particular, $H(\xi) < 0$ implies $H(\xi_M) < 0$ for large M. Similarly for below.

(4.8) **Lemma.** *Assume* $(r-2)^2 \ge 4\lambda_v d^2/c^2$. *If* $(DC)_1$ *holds, then there exists a variation ξ with $H(\xi) < 0$; so α_0 is unstable.*

(4.9) **Lemma.** *Assume* $(r-2)^2 \ge 4\lambda_v d^2/c^2$ *and*

(4.10) $\qquad \sqrt{(p-2)^2 + 4\lambda_u} + \sqrt{(r-2)^2 - 4\lambda_v d^2/c^2} > p + r - 4.$

Then there exists $\delta > 0$ such that

$$H(\xi) \ge \delta \|\xi\|^2$$

for every $\xi \in \mathcal{A}$; so α_0 is stable.

Proofs. We prove Lemmata (4.8) and (4.9) together. For that we consider the following eigenvalue problem:

Set

$$\sigma_0 = \inf \left\{ H(\xi) : \xi \in \mathcal{A} \quad \text{and} \quad \int_0^{\pi/2} \xi^2 \nu \, ds = 1 \right\}.$$

Formally, if $\sigma_0 < \infty$ and is achieved for some $\xi_0 \in \mathcal{A}$, then ξ_0 is a solution of the divergence equation

(4.11) $\qquad -\dfrac{d}{ds}(\dot{\xi}\nu) + L\,\xi\,\nu = \sigma\,\xi\,\nu,$

where $L = \dfrac{\lambda_u}{\sin^2} - \dfrac{d^2\lambda_v}{c^2\cos^2}$, and $\sigma \in \mathbb{R}$ appears as a Lagrange multiplier. We shall determine σ_0 by solving (4.11) explicitly.

Step 1. Take $\xi_0 = \sin^t \cos^{-q}$ with

$$t = \frac{1}{2}\left[\sqrt{(p-2)^2 + 4\lambda_u} - (p-2)\right]$$
$$q = \frac{1}{2}\left[(r-2) - \sqrt{(r-2)^2 - 4\lambda_v d^2/c^2}\right].$$

A straightforward but tedious computation shows that ξ_0 solves (4.11) with

(4.12) $\quad \sigma = \sqrt{(p-2)^2 + 4\lambda_u} + \sqrt{(r-2)^2 - 4\lambda_v d^2/c^2} - p - r + 4.$

Next, take $\xi = \xi_0$ in (4.11), multiply both sides by ξ_0, and integrate over $[0, \pi/2]$; we get

$$-\int_0^{\pi/2} \left[\frac{d}{ds}(\dot{\xi}_0 \, \nu)\right] \xi_0 \, ds + \int_0^{\pi/2} L\,\xi_0^2\,\nu\,ds = \sigma \int_0^{\pi/2} \xi_0^2\,\nu\,ds.$$

Integrate the first integral by parts to obtain

(4.13) $\quad H(\xi_0) = \sigma \int_0^{\pi/2} \xi_0^2 \, \nu \, ds.$

Because (ii) of (DC)$_1$ is just the condition $\sigma < 0$, we have verified Lemma (4.8).

Step 2. Now $\sigma > 0$. We prove

(4.14) $\quad \sigma = \sigma_0.$

On one hand, (4.13) implies $\sigma \geq \sigma_0$. For the converse, take a small $\varepsilon > 0$. Let $\mathcal{A}_\varepsilon$ be the subspace of $\mathcal{A}$ consisting of functions with support in $[\varepsilon, \pi/2 - \varepsilon]$. Set

$$\sigma_\varepsilon = \inf\left\{H(\xi) : \xi \in \mathcal{A}_\varepsilon \text{ and } \int_0^{\pi/2} \xi^2 \nu \, ds = 1\right\}.$$

Because L is bounded on $[\varepsilon, \pi/2 - \varepsilon]$, it is easy to prove that σ_ε is achieved by some $\xi_\varepsilon \in \mathcal{A}_\varepsilon$ satisfying

(4.15) $\quad -\frac{d}{ds}(\dot{\xi}_\varepsilon \, \nu) + L\,\xi_\varepsilon \, \nu = \sigma_\varepsilon \, \xi_\varepsilon \, \nu$

with $\xi_\varepsilon > 0$ on $(\varepsilon, \pi/2 - \varepsilon)$ and $\xi_\varepsilon(\varepsilon) = 0 = \xi_\varepsilon(\pi/2 - \varepsilon)$.

We claim $\sigma_\varepsilon > \sigma$. Indeed, multiplying (4.11) (with $\xi = \xi_0$) and (4.15) by ξ_ε and ξ_0 respectively and integrating by parts leads to

$$(\sigma_\varepsilon - \sigma) \int_\varepsilon^{\pi/2-\varepsilon} \xi_0 \, \xi_\varepsilon \, \nu \, ds = \dot{\xi}_0 \, \xi_\varepsilon \, \nu \Big|_{\pi/2-\varepsilon}^{\varepsilon} > 0.$$

164

That is because $\xi_\varepsilon > 0$ on $(\varepsilon, \pi/2 - \varepsilon)$ and 0 at the endpoints; hence $\dot\xi_\varepsilon(\varepsilon) > 0$ and $\dot\xi_\varepsilon(\pi/2 - \varepsilon) < 0$. But the integral on the left is positive, so $\sigma_\varepsilon > \sigma$.

Next,

(4.16) $$\sigma_\varepsilon \to \sigma_0 \quad \text{as} \quad \varepsilon \to 0.$$

For that we notice that $\mathcal{A}_\varepsilon \subset \mathcal{A}_{\varepsilon'} \subset \mathcal{A}$ for $\varepsilon > \varepsilon' > 0$. Therefore σ_ε is an increasing function of ε, and $\sigma_\varepsilon \geq \sigma_0$. Therefore,

$$\lim_{\varepsilon \to 0} \sigma_\varepsilon \geq \sigma_0.$$

On the other hand, for $\beta \in \mathcal{A}$ with $\int_0^{\pi/2} \beta^2 \nu \, ds = 1$, let $\beta_\varepsilon = \eta_\varepsilon \beta$, where η_ε is a continuous function with $\eta_\varepsilon \equiv 0$ on $[0, \varepsilon]$ and $[\pi/2 - \varepsilon, \pi/2]$, $\eta_\varepsilon \equiv 1$ on $[2\varepsilon, \pi/2 - 2\varepsilon]$; and η_ε is linear elsewhere. Then $\beta_\varepsilon \in \mathcal{A}_\varepsilon$; moreover, $\beta_\varepsilon \to \beta$ in $\mathcal{A}$ and $H(\beta_\varepsilon) \to H(\beta)$ as $\varepsilon \to 0$. But

$$H(\beta_\varepsilon) \geq \sigma_\varepsilon \int_0^{\pi/2} \beta_\varepsilon^2 \nu \, ds.$$

Letting $\varepsilon \to 0$ we obtain $H(\beta) \geq \lim_{\varepsilon \to 0} \sigma_\varepsilon$. Because β was chosen arbitrarily, we find $\sigma_0 \geq \lim_{\varepsilon \to 0} \sigma_\varepsilon$, proving (4.16).

Letting $\varepsilon \to 0$ in the inequality $\sigma_\varepsilon > \sigma$ we obtain $\sigma_0 \geq \sigma$, completing Step 2.

Step 3. The potential $L = \dfrac{\lambda_u}{\sin^2} - \dfrac{d^2 \lambda_v}{c^2 \cos^2}$ depends continuously on $\gamma = (\lambda_u, \tilde\lambda_v)$, where $\tilde\lambda_v = d^2 \lambda_v / c^2$. Let L^* be the potential corresponding to $\gamma^* = \gamma/(1-\delta)$; and σ_0^* the corresponding eigenvalue. Here $\delta > 0$ is a small constant. We can take σ_0^* arbitrarily close to σ_0. Then

$$H(\xi) = \delta \int_0^{\pi/2} \dot\xi^2 \nu \, ds + (1-\delta) \int_0^{\pi/2} (\dot\xi^2 + L^* \xi^2) \nu \, ds$$
$$\geq \delta \int_0^{\pi/2} \dot\xi^2 \nu \, ds + (1-\delta)\sigma_0^* \int_0^{\pi/2} \xi^2 \nu \, ds \geq \delta \|\xi\|^2,$$

thereby proving Lemma (4.9). //

(4.17) **Proposition.** *Assume $J(\alpha_{\pi/2}) \geq J(\alpha_0)$. Then α_0 is unstable iff* (DC)$_1$ *is satisfied.*

Proof. This is an application of Lemmata (4.4), (4.8), (4.9), together with the following observation (as in the proof of Proposition (3.12)): the condition that there exists $\xi \in \mathcal{A}$ with $H(\xi) < 0$ is *open*; i.e., it is preserved by small perturbations of $p, r, \lambda_u, \lambda_v, c, d$. Therefore, in the separating case

$$\sqrt{(p-2)^2 + 4\lambda_u} + \sqrt{(r-2)^2 - 4\lambda_v d^2/c^2} = p + r - 4,$$

α_0 must be stable; for otherwise a small perturbation would violate Lemma (4.9).
//

(4.18) Finally we prove Theorem (4.1):

By Propositions (3.12) and (4.17) we see that if (4.3) is satisfied, then there exists an equivariant harmonic join iff $(DC)_1$ holds. Now we prove that (4.3) together with $(DC)_1$ implies $(DC)_2$. Suppose otherwise:

$$\begin{cases} (p-2)^2 \geq 4\lambda_u c^2/d^2 \\ \text{and} \\ \sqrt{(r-2)^2 + 4\lambda_v} + \sqrt{(p-2)^2 - 4\lambda_u c^2/d^2} \geq p + r - 4. \end{cases}$$

After squaring the second inequality we conclude that either

(4.19) $$2\lambda_v - 2\lambda_u c^2/d^2 > (p-2)(r-2)$$

or (squaring again)

(4.20) $$\lambda_v + \lambda_u c^2/d^2 \leq \left((p-2)\lambda_v - (r-2)\lambda_u c^2/d^2\right)(p+r-4).$$

We have a contradiction: on one hand, (4.20) is not possible by (4.3). And from (4.19) we obtain

$$(p-2)^2(r-2) + 2(p-2)\lambda_u c^2/d^2 < 2(p-2)\lambda_v.$$

And using the assumption $(p-2)^2 \geq 4\lambda_u c^2/d^2$, we find

$$(2(r-2) + p - 2)\lambda_u c^2/d^2 < (p-2)\lambda_v;$$

thus

$$(r-2)\lambda_u c^2/d^2 < (p-2)\lambda_v,$$

again contradicting (4.3). Therefore $(DC)_2$ holds.

Finally, if (4.3) does not hold then $(DC)_2$ is necessary and sufficient for the existence of an equivariant harmonic join. That follows by interchanging the roles of α_0 and

$\alpha_{\pi/2}$ in Propositions (3.12) and (4.17). Furthermore, if (DC)$_2$ holds and (4.3) does not, then (DC)$_1$ is satisfied. //

5. EXAMPLES OF HARMONIC MAPS

(5.1) Inspection of (DC)$_1$, (DC)$_2$ of Theorem (4.1) reveals:

If $p = 2$, then (DC)$_2$ holds for every $d/c \in (0, +\infty)$.

If $p > 2$, there is $\varepsilon_2 > 0$ for which (DC)$_2$ holds iff $d/c \in (0, \varepsilon_2)$.

If $r = 2$, then (DC)$_1$ holds for every $d/c \in (0, +\infty)$.

If $r > 2$, there is $\varepsilon_1 > 0$ for which (DC)$_1$ holds iff $d/c \in (\varepsilon_1, \infty)$.

It is worth remarking that $(\varepsilon_1, \varepsilon_2)$ is *never* empty; for it always contains $d/c = \sqrt{(r-2)\lambda_u/(p-2)\lambda_v}$; that is precisely the value of the dilatation to insure $J(\alpha_0) = J(\alpha_{\pi/2})$ (see Lemma (4.2)).

For that specific value of d/c, the proof of the existence of a harmonic join can be much simplified, for it can be deduced from the second part of Lemma (2.11) ([ER], Sec.6).

(5.2) Corresponding to the decomposition $n = p + r - 1, p = 2, r = n - 1$, we have

(5.3) **Corollary.** *If $n \geq 3$ and $d^2/c^2 > (n-3)^2/4(n-2)$, then any map $\varphi : S^n \to Q^n(c,d)$ is homotopic to a harmonic map.*

Note that the restrictions are independent of the degree of φ.

Proof. If we take the identity map $v = Id : S^{r-1} \to S^{r-1}$ in the join (1.6), then $u *_\alpha v$ is homotopic to the rth suspension of u. Furthermore, each homotopy class of maps $\varphi : S^n \to Q^n(c,d)$ is represented by the $(n-1)^{st}$ suspension of the eigenmap $u = \varphi_k : S^1 \to S^1$ given by $\varphi_k(z) = z^k$ (VIII (1.25)), where k is the Brouwer degree of φ. We have $\lambda_u = k^2, \lambda_v = n - 2$. Then the Corollary follows from inspection of (DC)$_1$, (DC)$_2$. //

More generally, (5.1) tells us

(5.4) **Corollary.** *The homotopy class of the join $u * v : S^{p+r-1} \to Q^{q+s-1}(c,d)$ of two eigenmaps $u : S^{p-1} \to S^{q-1}, v : S^{r-1} \to S^{s-1}$ has a harmonic representative provided that the dilatation d/c is suitably restricted. In particular, for certain*

d/c the rth suspension of any eigenmap $u : S^{p-1} \to S^{q-1}$ can be harmonically represented ($r \geq 1$).

(5.5) Remark. The case $r = 1$ is not included in Theorem (4.1). However, it can be treated by similar methods: the explicit (DC) have been determined in [ER], Theorem (9.7).

(5.6) It is of interest to analyze the consequences of $(DC)_1$, $(DC)_2$ when $c = 1 = d$; i.e., the case of harmonic maps between Euclidean spheres. A detailed discussion is given in [PR].

Here are the key points: For $n \leq 7$ we can take $c = 1 = d$ in Corollary (5.3) to recover Smith's theorem [Sm2]. Corollary (5.3) does not include the case $c = 1 = d$ and $n = 9$. However, Cartan's harmonic eiconal $P : \mathbb{R}^8 \to \mathbb{R}$ of polynomial degree $k = 6$ (VIII (1.24)) has gradient $\nabla P = v : S^7 \to S^7$, an eigenmap of Brouwer degree 1; thus $\lambda_v = 55$, and inspection of $(DC)_1$, $(DC)_2$ with $c = 1 = d$ shows that v can be harmonically joined with any eigenmap $u = \varphi_k : S^1 \to S^1$. That gives harmonic maps $S^9 \to S^9$ of any Brouwer degree [EL4].

(5.7) Corollary. *If $u : S^{p-1} \to S^{q-1}$ and $v : S^{r-1} \to S^{s-1}$ are eigenmaps of the same polynomial degree, then they have an equivariant harmonic join $u *_\alpha v : S^{p+r-1} \to S^{q+s-1}$.*

(5.8) Corollary. *Let $u : S^{p-1} \to S^{q-1}$ be an eigenmap of polynomial degree ≥ 2; and $v = Id : S^{r-1} \to S^{r-1}$. Then the rth suspension of u can be represented by an equivariant harmonic map $u *_\alpha v : S^{p+r-1} \to S^{q+r-1}$ iff $1 \leq r \leq 6$.*

(5.9) Corollary (5.8) applies to each of the three Hopf fibrations: *The generators of each of the following homotopy groups have harmonic representatives:*

(i) $\pi_{n+1}(S^n) = \mathbb{Z}_2$ ($3 \leq n \leq 8$);

(ii) $\pi_{n+3}(S^n) = \mathbb{Z}_{24}$ ($5 \leq n \leq 10$);

(iii) $\pi_{n+7}(S^n) = \mathbb{Z}_{240}$ ($9 \leq n \leq 14$).

(5.10) The restriction $r \leq 6$ in Corollary (5.8) confines the equivariant harmonic join (of maps between Euclidean spheres) to low dimensions; in fact, with our present knowledge of eigenmaps, this method does not produce any non–contractible harmonic maps $S^m \to S^n$ for $n \geq 51$ (apart from $\pm Id : S^n \to S^n$). By way of contrast, in the next Chapter we shall show that the Hopf construction gives rise to non–contractible harmonic maps between Euclidean spheres of arbitrarily large dimensions.

(5.10) With reference to Corollary (5.8) we observe that not even eigenmaps of polynomial degree 2 can be suspended 7 times, because (ii) of $(DC)_1$ is violated. However, in those cases *arbitrarily small ellipsoidal deformations can produce harmonic maps*.

(5.11) **Remark.** There is (VIII (1.23)) an eigenmap $u : S^7 \to S^7$ of polynomial and Brouwer degree 2. Its equivariant harmonic 1st suspension represents $2 \in \pi_8(S^8) = \mathbb{Z}$. However, that class cannot be represented harmonically by an equivariant suspension of $\varphi_2 : S^1 \to S^1$, as noted in (5.10).

6. Notes and comments

(6.1) The basic case $c = 1 = d$ of Theorem (4.1) is due to Ding [Di] and Pettinati–Ratto [PR], completing the work of Smith [Sm2]. See also Appendix 4.

(6.2) Other ellipsoidal deformations have been considered in [ER]; *in particular, with $Q^n(c,d)$ as in Corollary (5.3), any map $\varphi : Q^n(c,d) \to Q^n(c,d)$ can be deformed to a harmonic map*.

(6.3) The theory of Morse on closed convex sets of Banach spaces can be found in [Ch], [CE], [St].

(6.4) Proposition (3.1) was proved by Smith [Sm2] in the spherical case. The present extension to ellipsoids was proved in [ER], using ideas of [R2].

(6.5) Xin [X2] has shown that for all $m \geq 2$ every element of odd degree in $\pi_{2m+1}(S^{2m+1})$ can be represented by a harmonic map between Euclidean spheres. He starts with the Reduction Theorem IV (4.13), using projection maps $\rho, \sigma : S^{2m+1} \to [0, \pi/4]$ constructed from the isoparametric function $F : \mathbb{C}^{m+1} \to \mathbb{R}$ given by

$$F(z) = (|x|^2 - |y|^2)^2 + 4 <x,y>^2, \quad \text{where } z = x + iy.$$

The geometry of F has been thoroughly studied by Cartan [C3] when $m = 2$; and by Nomizu [N] in general.

From there, Xin's proof is an application of Ding's method, very much along the lines of Sections 2-4.

(6.6) The requirement in (1.6) that u, v be eigenmaps (resp., in X (1.2) below that f be a bi-eigenmap) is a necessary condition that $u *_\alpha v$ (resp., the Hopf construction on f) be harmonic.

(6.7) From (5.9) we see that the generator of $\pi_4(S^3) = \mathbb{Z}_2$ has a harmonic representative. From [BK, Proposition 4.1] we find that it can also be represented by a rational map $\varphi : S^4 \to S^3$; i.e., by the restriction of a map $\Phi : \mathbb{R}^5 \to \mathbb{R}^4$ whose components Φ^α are quotients P^α/Q^α of polynomials with $Q^\alpha|_{S^4} \neq 0$ ($1 \leq \alpha \leq 4$). By way of contrast, Wood's theorem VIII (3.10) implies that any polynomial map $S^4 \to S^3$ is constant.

(6.8) The connected sum $\mathbb{C}P^m \# \overline{\mathbb{C}P}^m$ of complex projective spaces ($m \geq 2$) can be endowed with an Einstein metric to produce a manifold of cohomogeneity one; such constructions were made by Page and Bérard Bergery. Vanderwinden [Va] has established the existence of a harmonic map $\mathbb{C}P^m \# \overline{\mathbb{C}P}^m \to S^{2m}$ of degree one. Her proof follows the pattern of this Chapter – in particular, beginning with assumptions permitting reduction to a 1-dimensional variational problem.

CHAPTER X. THE HARMONIC HOPF CONSTRUCTION

INTRODUCTION

In the context of the Hopf construction, we prove a theorem resembling Theorem IX (4.1) (see (1.6)). Perhaps its most interesting feature is that it gives non–contractible harmonic maps between Euclidean spheres of large dimensions (see (2.14), for instance); furthermore, its applications (see Section 2) provide interesting relations between our analytical methods and classical theories in topology and linear algebra (J–homomorphisms, quadratic forms, Clifford algebras).

In Section 3 we construct a family of harmonic morphisms from S^3 (with suitable ellipsoidal metrics) to S^2 (Theorem (3.13)).

All the existence results of this Chapter are obtained by direct analysis of the harmonicity equations (1.5), (3.5).

1. THE EXISTENCE THEOREM

(1.1) As in Chapter IX, we parametrize points on S^{p+q-1} by $(\sin s \cdot x, \cos s \cdot y)$, $s \in [0, \pi/2], x \in S^{p-1}, y \in S^{q-1}$. If $f : S^{p-1} \times S^{q-1} \to S^{n-1}$ is a bi–eigenmap with bi–eigenvalue (λ, μ) (see VIII (2.11)), the α–Hopf construction on f is the map $\varphi : S^{p+q-1} \to S^n$ given by

(1.2) $$\varphi(\sin s \cdot x, \cos s \cdot y) = (\sin \alpha(s) \cdot f(x,y), \cos \alpha(s)),$$

where we parametrize the points of $S^n \subset \mathbb{R}^n \times \mathbb{R}^1$ by $(\sin \gamma \cdot z, \cos \gamma)$ with $z \in S^{n-1}$ and $0 \le \gamma \le \pi$; and $\alpha : [0, \pi/2] \to [0, \pi]$ satisfies the boundary conditions

(1.3) $$\alpha(0) = 0, \quad \alpha(\pi/2) = \pi.$$

Note that the map in (1.2) is homotopic to that of VIII (2.4); in particular, it coincides with it if $\alpha(s) = 2s$.

Following the derivation of Theorem IX (1.11), we obtain the

(1.4) **Reduction Theorem.** *The α–Hopf construction on the bi–eigenmap* $f : S^{p-1} \times S^{q-1} \to S^{n-1}$ *is harmonic iff*

(1.5) $$\ddot{\alpha} + ((p-1)\cot - (q-1)\tan))\dot\alpha - \left(\frac{\lambda}{\sin^2} + \frac{\mu}{\cos^2}\right)\sin\alpha\cos\alpha = 0,$$

where α satisfies (1.3).

The basic existence theorem is

(1.6) Theorem. *If $f : S^{p-1} \times S^{q-1} \to S^{n-1}$ is a bi–eigenmap with bi–eigenvalue (λ, μ), then there is a harmonic α–Hopf construction $\varphi : S^{p+q-1} \to S^n$ on f provided that the Hopf damping conditions*

(HDC)
$$\begin{cases} (p-2)^2 > 4\lambda \quad \text{and} \quad (q-2)^2 > 4\mu\,; \\ \text{or} \\ p = q \quad \text{and} \quad \lambda = \mu \end{cases}$$

are satisfied.

Proof. Step 1. Make the transformation

(1.7) $$\beta(t) = \alpha(\tan^{-1}(e^t)) - \pi/2 \quad \text{for} \quad t \in \mathbf{R}.$$

Then (1.5) becomes

(1.8) $$\beta''(t) + D(t)\beta'(t) + G(t)\sin\beta(t)\cos\beta(t) = 0,$$

where
$$D(t) = \frac{(p-2)e^{-t} - (q-2)e^{t}}{e^t + e^{-t}}, \quad G(t) = \frac{\mu e^t + \lambda e^{-t}}{e^t + e^{-t}}.$$

The boundary conditions (1.3) are now

(1.9) $$\lim_{t \to \pm\infty} \beta(t) = \pm \pi/2.$$

Equation (1.8) describes the motion of a pendulum with variable damping D and gravity G. That latter is bounded above and below by positive constants; thus it is sensible to look for solutions satisfying (1.9). We also notice that when $D(t)$ is positive it acts like friction – and in particular, decreases the speed; but when $D(t)$ is negative it increases the speed.

Step 2. For $s \in \mathbf{R}$ and $b \in [0, +\infty)$ let $\beta(s, b)$ be the unique solution of (1.8) with initial data $\beta(s) = 0, \beta'(s) = b$. Let $\beta^+(s)$ denote the collection of $b \in (0, +\infty)$ such that $\beta(s, b)$ increases monotonically to $\pi/2$ in finite time as t increases from s to $+\infty$. Similarly, $\beta^-(s)$ denotes the collection of $b \in (0, +\infty)$ such that $\beta(s, b)$ decreases monotonically to $-\pi/2$ in finite time as t decreases from s to $-\infty$. Now $\beta^\pm(s) \neq \emptyset$, so we define

(1.10) $$A^\pm(s) = \inf \beta^\pm(s).$$

Analysis of the pendulum equation (1.8) shows

(1.11) *The functions $A^{\pm}(s)$ are continuous.*

Verification of that fact is given in Appendix 4(8).

Moreover,

(1.12) *If $A^+(s) > 0$, then the solution $\beta(s, A^+(s))$ increases asymptotically to $\pi/2$ as t increases from s to $+\infty$. Similarly, if $A^-(s) > 0$, then $\beta(s, A^-(s))$ decreases asymptotically to $-\pi/2$ as t decreases from s to $-\infty$.*

Step 3. Until Step 8 below we put aside the symmetric case $p = q$, $\lambda = \mu$. We can assume $p, q > 2$. Denote by t_0 the point for which $D(t_0) = 0$. Take $b_1, b_2 \in \beta^+(s)$.

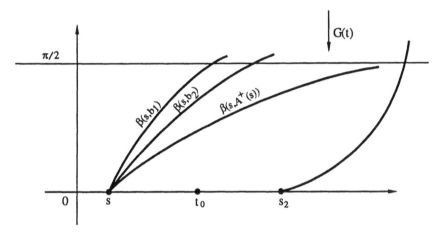

Figure 1.12

Think of $G(t)$ as a force pushing solutions toward zero; that suggests $A^+(s) > 0$, which would be so if $D(t)$ were positive. However, that is the case only for $s < t_0$. In fact, for $s > t_0$ it can happen that an arbitrarily small initial push takes the pendulum to $\pi/2$ in finite time, under the domination of damping. Balancing the effects of gravity and damping requires delicate estimates below; of course, these are the core of the proof.

Step 4.

(1.13) *If $s \leq t_0$, then $A^+(s) > 0$.*

(1.14) *There is a point $s_2 > t_0$ at which $A^+(s_2) = 0$.*

And similarly,

(1.13)′ *If $s \geq t_0$, then $A^-(s) > 0$.*

(1.14)′ *There is a point $s_1 < t_0$ at which $A^-(s_1) = 0$.*

In Step 5 below we shall prove (1.13) and (1.14); proofs of (1.13)′ and (1.14)′ are analogous. The idea for (1.13) is to construct explicit *subsolutions* with respect to which we can apply the following comparison principle:

(1.15) **Lemma.** *Assume there is a function F such that*

(i) $F(s) = 0, F'(s) > 0$;

(ii) *$F(t)$ increases asymptotically to $\pi/2$ as t increases from s to $+\infty$;*

(iii) *$F''(t) + D(t)F'(t) + G(t)\sin F(t)\cos F(t) > 0$ for all $t \in (s, +\infty)$.*

Then $A^+(s) \geq F'(s)$.

We give the proof in Appendix 4 – an application of a comparison theorem of Sturm ([CL],p.210). The main idea is the following: if F satisfies the conditions in Lemma (1.15), then for a suitable gravity $\tilde{G}(t) < G(t)$. with $t \in (s, +\infty)$, we have

$$F''(t) + D(t)F'(t) + \tilde{G}(t)\sin F(t)\cos F(t) = 0$$

on $(s, +\infty)$. Therefore the solution $\beta(s, F'(s))$ of (1.8) cannot reach $\pi/2$, because $G(t)$ pushes it toward zero in a stronger way than $\tilde{G}(t)$ ($< G(t)$) does for F; thus $A^+(s) \geq F'(s)$. //

Step 5. To prove (1.13) we shall construct subsolutions as in Lemma (1.15) for $s \leq t_0$.

We construct a subsolution F explicitly for $s = t_0$ by taking

(1.16) $$F(t) = F_c(t) = \tan^{-1}(c(t - t_0)),$$

where $c > 0$ is a constant to be determined. Clearly F_c satisfies (i) and (ii) of Lemma (1.15). To check (iii) we substitute F_c, and use the identity

$$\sin x \cos x = \tan x/(1 + \tan^2 x)$$

to obtain, up to the positive factor $c(t - t_0)/(1 + c^2(t - t_0)^2)$

(1.17) $$\frac{-2c^2}{1 + c^2(t - t_0)^2} + \frac{D(t)}{t - t_0} + G(t).$$

With $y = t - t_0$, that becomes

(1.18) $$\frac{-2c^2}{1 + c^2 y^2} + \frac{R(y)}{Q(y)},$$

where

$$R(y) = \sqrt{(p-2)(q-2)} \, \frac{e^{-y} - e^{y}}{y} + \mu \sqrt{\frac{p-2}{q-2}} \, e^{y} + \lambda \sqrt{\frac{q-2}{p-2}} \, e^{-y}$$

and

$$Q(y) = \sqrt{\frac{p-2}{q-2}} \, e^{y} + \sqrt{\frac{q-2}{p-2}} \, e^{-y} \, .$$

Next, we observe that $\lambda \geq p - 1 > p - 2$ and $\mu \geq q - 1 > q - 2$, because λ (resp., μ) is an eigenvalue of $\Delta^{S^{p-1}}$ (resp., $\Delta^{S^{q-1}}$).

Thus

(1.19) $\quad R(y) > \sqrt{(p-2)(q-2)} \left[\frac{e^{-y} - e^{y}}{y} + e^{y} + e^{-y} \right] > 0$

for all $y \in (0, +\infty)$, because $\tanh y < y$. Moreover

(1.20) $\quad \lim_{y \to 0} R(y) = \mu \sqrt{\frac{p-2}{q-2}} + \lambda \sqrt{\frac{q-2}{p-2}} - 2\sqrt{(p-2)(q-2)} > 0 \, ;$

and

(1.21) $\quad \lim_{y \to +\infty} \frac{R(y)}{Q(y)} = \mu \, .$

From (1.19), (1.20), (1.21) we conclude that $R(y)/Q(y)$ is positive and bounded away from 0 on $(0, +\infty)$. Thus inspection reveals that the expression (1.18) is strictly positive on $(0, +\infty)$, provided that $c = \bar{c}$ is sufficiently small. Therefore $F_{\bar{c}}(t)$ satisfies (i), (ii), (iii).

The damping force $D(t)$ is friction for $t < t_0$; thus a simple modification of (1.16) yields the required subsolutions when $s < t_0$.

Step 6. We prove (1.14). That requires hypothesis (HDC); and we shall use $(q - 2)^2 > 4\mu$. Suppose that $A^{+}(s) > 0$, and let $y_s(t)$ be the solution $\beta(s, A^{+}(s))$; we can write

(1.22) $\quad y_s(t) = \tan^{-1} \left[\exp \left(\int_{\bar{t}}^{t} H(u) du + d \right) \right]$

for $t \in (s, +\infty)$ and for suitable constants $\bar{t} \in (s, +\infty)$ and $d \in \mathbb{R}$; the function $H(u)$ is uniquely determined by $y_s(t)$.

(1.23) **Lemma.** (*Asymptotic estimates.*)

(1.24)
$$\lim_{t \to s^+} H(t) = +\infty,$$

(1.25)
$$\lim_{t \to +\infty} H(t) = m,$$

where $m > 0$ is determined through $\mu = m(m + q - 2)$.

Proof. (1.24) is elementary. Just observe that

$$\lim_{t \to s^+} H(t) = \lim_{t \to s^+} \frac{y_s'(t)}{y_s(t)} = +\infty.$$

For (1.25), note that

$$\lim_{t \to +\infty} G(t) = \mu, \quad \lim_{t \to +\infty} D(t) = -(q-2).$$

Also, since $\lim_{t \to +\infty} y_s(t) = \pi/2$, we have

$$\sin y_s(t) \cos y_s(t) \sim \pi/2 - y_s(t) \quad \text{for } t \text{ large}.$$

In conclusion, the harmonicity equation is asymptotically approximated by

(1.26)
$$\begin{cases} w_s''(t) - (q-2)w_s'(t) + \mu(\pi/2 - w_s(t)) = 0, \\ \lim_{t \to +\infty} w_s = \pi/2. \end{cases}$$

That implies

$$y_s(t) \sim w_s(t) = \pi/2 - ce^{-mt} \quad (c \in \mathbb{R})$$

for t large, from which (1.25) follows easily. //

To simplify notation we set

$$f(t) = \exp\left(\int_t^t H(u)du + d\right).$$

Since $y_s(t)$ satisfies (1.8), we can substitute (1.22) in (1.8) to obtain

$$\frac{f''(t)}{f(t)} - \frac{2f'^2(t)}{1+f^2(t)} + D(t)\frac{f'(t)}{f(t)} + G(t) = 0.$$

Therefore

(1.27)
$$\frac{f''(t)}{f(t)} + D(t)\frac{f'(t)}{f(t)} + G(t) > 0 \quad \text{for } t \in (s, +\infty),$$

which is equivalent to

(1.28) $\quad H'(t) + H^2(t) + D(t)H(t) + G(t) > 0 \quad \text{for} \quad t \in (s, +\infty).$

Define the quadratic form

$$V_t(x) = x^2 + D(t)x + G(t).$$

If a_t, b_t are its roots, then $V_t(x) \leq 0$ when they are real and $x \in [a_t, b_t]$. As t ranges over $(s, +\infty)$, the intervals $[a_t, b_t]$ sweep out an area (possibly empty) which we denote by HND in Figure 1.28 below; if $H(t) \in [a_t, b_t]$, then $H'(t) > 0$ by (1.28).

Now, the hypothesis $(q - 2)^2 > 4\mu$ guarantees that HND $\neq \emptyset$ for t large; namely, simple computations show

(1.29)
$$\lim_{t \to +\infty} a_t = \frac{1}{2}\left[(q-2) - \sqrt{(q-2)^2 - 4\mu}\right]$$
$$\lim_{t \to +\infty} b_t = \frac{1}{2}\left[(q-2) + \sqrt{(q-2)^2 - 4\mu}\right].$$

Moreover, from $\mu = m(m + q - 2)$ we find

$$m = \frac{1}{2}\left[-(q-2) + \sqrt{(q-2)^2 + 4\mu}\right]$$

and therefore

(1.30) $\quad m < \frac{1}{2}\left[(q-2) - \sqrt{(q-2)^2 - 4\mu}\right].$

For otherwise, $\sqrt{(q-2)^2 + 4\mu} + \sqrt{(q-2)^2 - 4\mu} \geq 2(q-2)$, so by squaring $\sqrt{(q-2)^4 - 16\mu^2} \geq (q-2)^2$, which is not acceptable.

To complete the proof of (1.14), suppose $A^+(s) > 0$ (s large). Clearly (1.29) and (1.30) together with (1.24) and (1.25) force the function $H(t)$ associated with $y_s(t)$ to cross HND with $H'(t) < 0$ somewhere, which is not allowed. Therefore, there must be an s_2 satisfying (1.14).

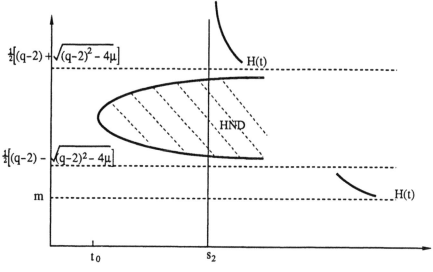

Figure 1.28

The proof of (1.14)' is similar, using $(p - 2)^2 > 4\lambda$.

Step 7. We prove Theorem (1.6) in the non-symmetric case: From (1.13), (1.14), (1.13)', (1.14)' together with (1.11) we conclude that there exists a point $\bar{s}$ with $A^-(\bar{s}) = A^+(\bar{s}) > 0$:

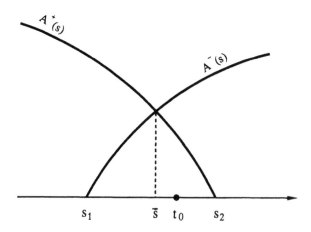

By (1.12) the solution $\beta(\bar{s}, A^-(\bar{s})) = \beta(\bar{s}, A^+(\bar{s}))$ satisfies (1.9), as required.

Step 8. In the symmetric case $p = q$, $\lambda = \mu$, the assumptions $(p-2)^2 > 4\lambda$, $(q-2)^2 > 4\mu$ are unnecessary. In fact, we need only $\lambda > p-2, \mu > q-2$ to make sure that $A^\pm(0) > 0$. Thus the symmetry forces $A^+(\bar{s}) = A^-(\bar{s})$ with $\bar{s} = 0$. //

(1.31) **Remark.** The hypothesis HDC in Theorem (1.6) cannot be entirely removed. In fact, ([Sm2],[R2]), *if*

$$\begin{cases} p = 2 = q & \text{and} \quad \lambda \neq \mu \\ \text{or} \\ p = 2, q > 2 & \text{and} \quad (q-1)\lambda \geq \mu \end{cases}$$

the harmonicity equation (1.5) does not admit solutions with the prescribed boundary values (1.3).

2. EXAMPLES OF HOPF'S CONSTRUCTION

An immediate application of Theorem (1.6) is

(2.1) **Theorem.** *Hopf's construction $\varphi : S^{p+q-1} \to S^n$ on an orthogonal multiplication $f : \mathbb{R}^p \times \mathbb{R}^q \to \mathbb{R}^n$ is homotopic to a harmonic map, provided that either $p = q$ or $p, q > 6$.*

When $p = q$, that was already established in VIII (2.5); in this case the solution of (1.5) is $\alpha(s) = 2s$. The case $q = n$ is of special interest, because of the Hurwitz–Radon multiplications (VIII (2.7)).

For the next applications, we need some topological preparations [A2].

(2.2) The natural inclusion map $\mathbb{R}^n \hookrightarrow \mathbb{R}^{n+1}$ induces an inclusion $O(n) \hookrightarrow O(n+1)$ of the associated orthogonal groups. We denote by

$$O(\infty) = \lim_{n \to \infty} O(n)$$

the limit group.

Similarly, let

$$\pi_p = \lim_{n \to \infty} \pi_{p+n}(S^n)$$

following the suspension homomorphism $\pi_{p+n}(S^n) \to \pi_{p+n+1}(S^{n+1})$, defined for all p, n.

(2.3) The *J–homomorphism*

$$J_p^n = J_p : \pi_p(O(n)) \to \pi_{p+n}(S^n)$$

is defined through the following construction: for any map $\gamma : S^p \to O(n)$ let $f_\gamma : S^p \times S^{n-1} \to S^{n-1}$ be given by

(2.4) $$f_\gamma(x, y) = \gamma(x) \cdot y .$$

Denote by $J_p(\gamma) : S^{p+n} \to S^n$ Hopf's construction of f_γ. J_p induces a *homomorphism* on homotopy groups, which is *stable* for $n - 1 > p$: i.e., the following diagram

(2.5)
$$\begin{array}{ccc} \pi_p(O(n)) & \xrightarrow{J_p} & \pi_{p+n}(S^n) \\ \downarrow & & \downarrow \\ \pi_p(O(n+1)) & \xrightarrow{J_p} & \pi_{p+n+1}(S^{n+1}) \end{array}$$

is commutative and the vertical arrows (described in (2.2)) are isomorphisms. Letting $n \to +\infty$ in (2.5) defines

(2.6) $$J_p : \pi_p(O(\infty)) \to \pi_p .$$

J_p in (2.6) is called the *stable J–homomorphism*.

Its basic properties are the following:

(2.7) J_p *is a monomorphism for* $p = 0, 1$ mod 8. *Furthermore, in this case* $Im\ J_p = \mathbb{Z}_2$.

(2.8) If $p = 3, 7$ mod 8, write $p = 4s - 1$ and $m(2s) = $ denominator of $(-1)^{s-1}B_s/4s$, where B_s is the sth Bernoulli number: here the fraction $B_s/4s$ is expressed in lowest terms. Then $Im(J_{4s-1}) = \mathbb{Z}_{m(2s)}$. Moreover, $Im(J_p)$ is a direct summand in π_p.

(2.9) For $p = 2, 4, 5, 6$ mod 8, $Im(J_p) = 0$; that is because $\pi_p(O(\infty)) = 0$ for those values of p – recalling that $\pi_p(O(\infty)) = \pi_{p+8}(O(\infty))$, by Bott's periodicity theorem.

(2.10) Atiyah, Bott, Shapiro [ABS] have shown how to represent the elements of $\pi_p(O(\infty))$ as equivalence classes of finitely generated C_p–modules (C_p = the indicated Clifford algebra). And Baum [B] has interpreted that by saying that every element in $Im\ J_p \subset \pi_p$ is represented by Hopf's construction on an orthogonal multiplication $f(p+1, n; n)$ for some n (see also [He]).

So we have the characterization:

(2.11) *$Im\ J_p$ consists precisely of those stable homotopy classes representable by Hopf's construction on an orthogonal multiplication of the type $f(p+1, n; n)$.*

We are now in position to apply Theorem (1.6) to obtain theorems on harmonicity.

(2.12) **Corollary.** *Any quadratic form $\varphi : S^m \to S^n$ is homotopic to a harmonic map, provided $m \geq n + 6$.*

Proof. Theorem VIII (3.19) asserts that φ is homotopic to some Hopf construction on an $f(p, q; n)$. Here $\lambda = p - 1, \mu = q - 1$. The damping conditions (HDC) are satisfied for $p, q > 6$, which is the case because $p + q - 1 = m \geq n + 6$ (and $p, q \leq n$). The Corollary now follows from Theorem (1.6). //

(2.13) **Corollary.** *Any element of π_p in $Im\ J_p$ has a harmonic representative, provided $p \geq 6$.*

More precisely, if $\gamma \in \pi_p(O(\infty))$, then for some integer n with $n - 1 > p \geq 6$ there is an $f_\gamma(p+1, n; n)$ for which $J_p^n(\gamma) \in \pi_{p+n}(S^n)$ has a harmonic representative.

Generators of $Im\ J_p$ occur when $c = 0$ in VIII (2.7); as an example of (2.13) we cite

(2.14) **Theorem.** *For each $i \geq 1$, set $p = 8i$ and $n = 2^{4i}$. Then there is a harmonic map $S^{p+n} \to S^n$ representing the generator of $Im\ J_p = \mathbb{Z}_2$.*

Similarly, we have essential harmonic maps associated with the other cyclic groups $Im\ J_p$ for $p = 1, 3, 7 \mod 8$.

(2.15) There is a generalized J–homomorphism

$$J : \pi_p(V_{n,k}) \to \pi_{p+k}(S^n).$$

Hopf's construction on orthogonal multiplications $f(p+1, k; n)$ produces representatives of classes in the image of this J, as well.

(2.16) All Hopf constructions on an $f(p, q; n)$ are even; thus we have harmonic maps $\mathbb{R}P^{p+q-1} \to S^n$, for $p = q$ or $p, q > 6$.

3. $\pi_3(S^2)$ AND HARMONIC MORPHISMS

(3.1) As in Chapter V (2.12), we consider the ellipsoid

$$Q^3(a,b) = \{x,y) \in \mathbb{C} \times \mathbb{C} : |x|^2/a^2 + |y|^2/b^2 = 1\}.$$

For suitable dilatations b/a we shall construct harmonic maps

$$\varphi_{k,\ell} : Q^3(a,b) \to S^2$$

with prescribed Hopf invariant $k.\ell$. These maps have the special feature of satisfying an integrability condition – horizontal conformality, as in IV (2.1) – which makes them *harmonic morphisms*.

(3.2) We start with the multiplication

$$f_{k,\ell} : S^1 \times S^1 \to S^1$$

defined by $f_{k,\ell}(e^{i\xi}, e^{i\eta}) = e^{i(k\xi + \ell\eta)}$ where $k, \ell \in \mathbb{Z}$ and $0 \leq \xi, \eta < 2\pi$. The map $f_{k,\ell}$ is a bi-eigenmap with bi-eigenvalue (k^2, ℓ^2). We apply the α-Hopf construction to it, defining

(3.3) $$\varphi_{k,\ell} : Q^3(a,b) \to S^2 \subset \mathbb{R}^2 \times \mathbb{R}^1$$

by

$$(a \sin s \cdot e^{i\xi}, b \cos s \cdot e^{i\eta}) \to (\sin \alpha(s) f_{k,\ell}(e^{i\xi}, e^{i\eta}), \cos \alpha(s))$$

where $\alpha : [0, \pi/2] \to [0, \pi]$ satisfies the boundary conditions

(3.4) $$\alpha(0) = 0, \quad \alpha(\pi/2) = \pi.$$

$\varphi_{k,\ell}$ *has Hopf invariant $k.\ell$; i.e., it represents $k.\ell \in \mathbb{Z} = \pi_3(S^2)$.*

Note also that for $k = 1 = \ell, a = 1 = b$ and $\alpha(s) = 2s$, then (3.3) is the Hopf fibration $S^3 \to S^2$ (VIII (2.8)). That is (up to adjustment of radii) a Riemannian submersion, and therefore a harmonic morphism.

A computation similar to Theorem (1.4) shows that a map (3.3) is harmonic iff

(3.5) $$\ddot{\alpha} + (\cot - \tan)\dot{\alpha} - \frac{\dot{h}}{h}\dot{\alpha} - h^2 \left(\frac{k^2}{a^2 \sin^2} + \frac{\ell^2}{b^2 \cos^2} \right) \sin \alpha \cos \alpha = 0.$$

where $h^2(s) = a^2 \cos^2 s + b^2 \sin^2 s$.

(3.6) We recall from IV (2.1) that a map $\varphi : M \to N$ is *horizontally conformal* if at each point $x \in M$, either $d\varphi(x) = 0$ or it is a conformal epimorphism. *Harmonic morphisms are characterized as the harmonic horizontally conformal maps.*

(3.7) Let us work out explicitly the condition of horizontal conformality of $\varphi_{k,\ell}$: Consider the open subset of $Q^3(a, b)$ parametrized by $0 \leq \xi, \eta < 2\pi, 0 < s < \pi/2$. The Riemannian metric g of $Q^3(a, b)$ is $g = a^2 \sin^2 s \, d\xi^2 + b^2 \cos^2 s \, d\eta^2 + h^2(s) ds^2$. We construct the orthonormal base

(3.8) $\quad e_1 = \dfrac{1}{a \sin s} \dfrac{\partial}{\partial \xi}, \quad e_2 = \dfrac{1}{b \cos s} \dfrac{\partial}{\partial \eta}, \quad e_3 = \dfrac{1}{h(s)} \dfrac{\partial}{\partial s}$

for the tangent space $T_{(\xi, \eta, s)}(Q^3(a, b))$.

Similarly, on S^2 we use coordinates $0 \leq \gamma < 2\pi, 0 \leq t \leq \pi$. Its Riemannian metric is $\sin^2 t \, d\gamma^2 + dt^2$. Thus the vectors

(3.9) $\qquad\qquad \dfrac{1}{\sin t} \dfrac{\partial}{\partial \gamma}, \quad \dfrac{\partial}{\partial t}$

form an orthonormal base of $T_{(\gamma, t)}(S^2), 0 < t < \pi$.

The differential of the map $\varphi_{k,\ell} = \varphi$ is

(3.10) $\quad \begin{cases} d\varphi(e_1) = \dfrac{k}{a \sin s} \dfrac{\partial}{\partial \gamma} \\ d\varphi(e_2) = \dfrac{\ell}{b \cos s} \dfrac{\partial}{\partial \gamma} \\ d\varphi(e_3) = \dfrac{\dot\alpha(s)}{h(s)} \dfrac{\partial}{\partial t} \end{cases}$

Now assume $b/a = \ell/k$; it follows that $\operatorname{Ker} d\varphi$ is spanned by the vector $v = \sin s \cdot e_1 - \cos s \cdot e_2$; and so an orthonormal base for its orthogonal complement is (e_3, w), where

$$w = \cos s \cdot e_1 + \sin s \cdot e_2 .$$

It is clear from (3.9) and (3.10) that $d\varphi$ preserves the orthogonality between e_3 and w. Thus *horizontal conformality* is expressed by

(3.11) $\qquad\qquad \|d\varphi(e_3)\|^2 = \|d\varphi(w)\|^2 .$

The case $b/a = -\ell/k$ can be handled similarly. Indeed, when $b/a = |\ell/k|$ (3.11) takes the form:

(3.12) $\qquad \dfrac{\dot\alpha^2(s)}{h^2(s)} = \left(\dfrac{|k| \cos s}{a \sin s} + \dfrac{|\ell| \sin s}{b \cos s} \right)^2 \sin^2 \alpha(s) .$

With these preparations we can state

(3.13) Theorem. *For any $k, \ell \in \mathbb{Z}$ there is an equivariant harmonic map*

$$\varphi_{k,\ell} : Q^3(a,b) \to S^2$$

as in (3.3) iff

(3.14)
$$b/a = |\ell/k|.$$

Furthermore, $\varphi_{k,\ell}$ is a harmonic morphism.

Proof. We change parameter, setting $\tan s = e^t$; and

$$A(t) = \alpha(\tan^{-1} e^t)$$
$$H(t) = h(\tan^{-1} e^t).$$

Then (3.5) takes the form

(3.15)
$$A'' = \frac{H'}{H} A' + GH^2 \sin A \cos A$$

where

(3.16)
$$G(t) = \frac{(\ell^2/b^2)e^t + (k^2/a^2)e^{-t}}{e^t + e^{-t}}, \quad H^2(t) = \frac{a^2 e^{-t} + b^2 e^t}{e^t + e^{-t}}.$$

The boundary conditions (3.4) become

(3.17)
$$\lim_{t \to -\infty} A(t) = 0, \quad \lim_{t \to +\infty} A(t) = \pi.$$

Step 1. (3.14) is necessary. Suppose that (3.15) has a solution A with limits (3.17). We argue as in the proof of Proposition IX (3.1) to find $A' > 0$. As an equation in H, (3.15) has the prime integral

(3.18)
$$H^2 = A'^2 / \left[2 \int_{-\infty}^{t} G \sin A \cos A\, A'\, dt \right].$$

We observe that for $|t|$ large, (3.15) behaves qualitatively like the equation of a pendulum with 0 damping and constant gravity. Therefore the assumption $\lim_{t \to +\infty} A(t) = \pi$ implies $\lim_{t \to +\infty} A'(t) = 0$. Together with (3.18), that forces

(3.19)
$$\int_{-\infty}^{+\infty} G \sin A \cos A\, A'\, dt = 0.$$

If (3.14) does not hold, then G is strictly monotone, which in turn requires that the integral in (3.19) not vanish.

Step 2. (3.14) is sufficient. We have now $G \equiv k^2/a^2 (= \ell^2/b^2) =$ constant. Take $\bar{t} \in \mathbb{R}$ and $0 < \varepsilon < \pi/2$; well-known properties of pendulum equations insure the existence of a solution $\bar{A}$ of (3.15) with

(3.20) $$\begin{cases} \bar{A}(\bar{t}) = \varepsilon; \lim_{t \to -\infty} \bar{A}(t) = 0; \\ \bar{A}'(t) > 0 \quad \text{for all} \quad t \in (-\infty, \bar{t}]. \end{cases}$$

We show that $\bar{A}$ must also satisfy $\lim_{t \to +\infty} \bar{A}(t) = \pi$. For that, we first observe that (3.18) is now

(3.21) $$\frac{\ell^2 e^t + k^2 e^{-t}}{e^t + e^{-t}} = \frac{\bar{A}'^2}{\sin^2 \bar{A}},$$

which holds on $(-\infty, \bar{t}]$. But $\bar{A}(t) \in (0, \pi)$ for all $t \in \mathbb{R}$; for if t_0 is the first point for which $\bar{A}(t_0) = 0$, then $\bar{A}'(t_0) = 0$ as well by (3.21), contradicting the uniqueness of solutions of (3.15). Similarly, $\bar{A}(t_0) = \pi$ is unacceptable. We conclude that $\bar{A}' > 0$ on $\mathbb{R}$, and so (3.21) holds globally. Therefore $\lim_{t \to +\infty} \bar{A}(t)$ exists and clearly equals π.

Moreover, it is easy to check that the transformation $\tan s = e^t$ carries the horizontal conformality condition (3.12) (with $|k/a| = |\ell/b|$) into (3.21).

The proof of Theorem (3.13) is completed by a routine verification that horizontal conformality is also valid across the loci $s = 0$ and $s = \pi/2$. //

(3.22) **Corollary.** *Let g be any metric on S^2, and $k, \ell \in \mathbb{Z}$. Then there is a harmonic morphism*
$$\psi_{k,\ell} : Q^3(a, b) \to (S^2, g)$$
with Hopf invariant $k \cdot \ell$, provided $\ell^2/k^2 = b^2/a^2$.

That is an immediate consequence of Theorem (3.13) and IV (2.3), for there is a conformal diffeomorphism $\gamma : S^2 \to (S^2, g)$, which is itself a harmonic morphism. //

In particular, composition of $\psi_{k,\ell}$ with any minimal immersion $i : (S^2, g = i^*h) \to (N, h)$ is harmonic.

(3.23) **Remark.** Let
$$C(\varphi_{k,\ell}) = \{x \in Q^3(a, b) : d\varphi_{k,\ell}(x) = 0\}.$$

(i) If $k = 1 = \ell$, then $C(\varphi_{k,\ell}) = \emptyset$;

(ii) If $k = 1, \ell \neq 1$, then $C(\varphi_{k,\ell})$ is the locus $s = \pi/2$;

(iii) If $k \neq 1, \ell = 1$, then $C(\varphi_{k,\ell})$ is the locus $s = 0$;

(iv) If $k \neq 1, \ell \neq 1$, then $C(\varphi_{k,\ell})$ is the union of the loci $s = 0$ and $s = \pi/2$.

(3.24) **Remark.** Since $\varphi_{k,\ell}$ is a harmonic morphism with 2–dimensional range, the fibre over any regular value is a minimal submanifold by IV (2.4). Thus each $\varphi_{k,\ell}$ determines an interesting family of closed geodesics on $Q^3(a, b)$.

Moreover, the pre–images of the longitudes are minimal surfaces in $Q^3(a, b)$. Compare that with the construction in Chapter V (2.12).

(3.25) **Remark.** The solution $\bar{A}$ to (3.15) with limits (3.17) is not unique. Others can be obtained by letting ε vary in (3.20). That produces variations of $\varphi_{k,\ell}$ through equivariant harmonic morphisms.

(3.26) **Remark.** Theorem (4.1) of Chapter IX proves the existence of harmonic maps for dilatations lying in suitable open intervals. By way of contrast, Theorem (3.13) produces such a map for a unique dilatation. That may reflect the fact that small perturbations of an integrable system are not generally integrable.

4. **Notes and comments**

(4.1) The case $p \neq q$ of Theorem (1.6) and its applications were obtained in [R1]. Functions $A^{\pm}$ similar to those used in the proof were introduced in [Sm1]. The transformation (1.22) was first used in [PR]. (See also Appendix 4.) Asymptotic estimates similar to those in (1.23) can be found in [Ha], pp.304–7.

(4.2) The harmonicity equation (1.5) is the Euler–Lagrange equation of the reduced energy functional

$$J(\alpha) = \int_0^{\pi/2} \left\{ \dot{\alpha}^2 + \left(\frac{\lambda}{\sin^2} + \frac{\mu}{\cos^2} \right) \sin^2 \alpha \right\} \sin^{p-1} \cos^{q-1} ds .$$

We note that our solutions in Theorem (1.6) are not absolute minima for J; these are the trivial solutions $\alpha \equiv 0, \alpha \equiv \pi$. Perhaps the existence of special solutions in Theorem (1.6) is related to stability properties of the constant solution $\alpha \equiv \pi/2$.

(4.3) The properties (2.7), (2.8) are essentially due to Adams [A2], supplemented by Quillen [Q].

(4.4) Theorem (3.13) is due to the authors [ER]. In case $a = 1 = b$ and Hopf invariant k^2, equivariant harmonic maps $\varphi : S^3 \to S^2$ were constructed by Smith [Sm1, 2]. On S^3 we can consider Riemannian metrics of the form $g = a^2 \sin^2 s \, d\xi^2 + b^2 \cos^2 s \, d\eta^2 + Y^2(s) \, ds^2$ [R3]. Then Theorem (3.13) holds with (S^3, g) in place of $Q^3(a, b)$. The proof is similar. In a series of papers (see [BW] and references therein) Baird and Wood have established a bijective correspondence between harmonic morphisms from a 3-dimensional manifold M to Riemann surfaces and conformal foliations of M by geodesics which make M a Seifert fibre space without reflections. In particular, they re-obtained Theorem (3.13) by a method which requires no explicit discussion of ordinary differential equations.

(4.5) Let G be a Lie group acting isometrically on M and N, and $\varphi : M \to N$ a G-equivariant map (i.e., $\varphi(g \cdot x) = g \cdot \varphi(x)$ for all $x \in M, g \in G$). The examples in Chapters V, VI, VII are G-equivariant. By way of contrast, most of the examples in Chapters IX, X are not G-equivariant – even if these are constructed using homogeneous isoparametric maps such as ρ, σ in IX (1.7).

APPENDIX 1.   SECOND VARIATIONS

E – index of a map

(1) For simplicity of exposition, we suppose that M is compact. Take a 2–parameter variation $\Phi : M \times \mathbf{R}^2 \to N$ of a harmonic map $\varphi = \Phi(\cdot, 0) : M \to N$, writing $\Phi(\cdot, s, t) = \varphi_{s,t}$, with

$$v = \left.\frac{\partial \varphi_{s,t}}{\partial s}\right|_{0,0} \qquad w = \left.\frac{\partial \varphi_{s,t}}{\partial t}\right|_{0,0}.$$

The E-Hessian of φ is the symmetric bilinear form on $C(\varphi^{-1}T(N))$ defined by

(2) $H_\varphi^E(v, w) = \left.\dfrac{\partial^2 E(\varphi_{s,t})}{\partial t \partial s}\right|_{0,0}$. Write $H_\varphi = H_\varphi^E$ for short. We calculate as in I (1.9):

$$\frac{\partial e(\Phi)}{\partial s} = \left\langle \nabla^\Phi \frac{\partial \Phi}{\partial s}, d\varphi_{s,t} \right\rangle,$$

where we view $d\varphi_{s,t}$ as the differential along $T(M)$. Then

$$\frac{\partial^2 e(\Phi)}{\partial t \partial s} = \left\langle \nabla_{\partial_t} \nabla^\Phi \frac{\partial \Phi}{\partial s}, d\varphi_{s,t} \right\rangle + \left\langle \nabla^\Phi \frac{\partial \Phi}{\partial s}, \nabla_{\partial_t} d\varphi_{s,t} \right\rangle.$$

As in Step 1 of I (1.9), we have

(3) $$\left\langle \nabla^\Phi \frac{\partial \Phi}{\partial s}, \nabla_{\partial_t} d\varphi_{s,t} \right\rangle\bigg|_{0,0} = \langle \nabla^\varphi v, \nabla^\varphi w \rangle.$$

For any vector field $X \in C(T(M))$ we have $[X, \partial_t] = 0$, so

$$\nabla_{\partial_t}^\Phi \nabla_X^\Phi = \nabla_X^\Phi \nabla_{\partial_t}^\Phi + R^\Phi(X, \partial_t),$$

where $R^\Phi(X, \partial_t) = R^N \left(d\Phi \cdot X, \dfrac{\partial \Phi}{\partial t} \right)$. We use the sign conventions on curvature in [EL1,2,3]; thus $R(X, Y)W = \nabla_{[X,Y]} W - [\nabla_X, \nabla_Y]W$. Therefore

(4) $$\left\langle \nabla_{\partial_t} \nabla^\Phi \frac{\partial \Phi}{\partial s}, d\varphi_{s,t} \right\rangle =$$
$$\left\langle \nabla^\Phi \nabla_{\partial_t}^\Phi \frac{\partial \Phi}{\partial s}, d\varphi_{s,t} \right\rangle + \left\langle R^N \left(d\varphi_{s,t}, \frac{\partial \Phi}{\partial t} \right) \frac{\partial \Phi}{\partial s}, d\varphi_{s,t} \right\rangle.$$

Viewing $\nabla^{\Phi}_{\partial_t} \dfrac{\partial \Phi}{\partial s}$ as a section of $\varphi^{-1}_{s,t} T(N)$, we have $\nabla^{\Phi} \nabla^{\Phi}_{\partial_t} \dfrac{\partial \Phi}{\partial s} = d \nabla^{\Phi}_{\partial_t} \dfrac{\partial \Phi}{\partial s}$.
Therefore,

$$\dfrac{\partial^2 e(\Phi)}{\partial t \partial s} = \left\langle d\nabla^{\Phi}_{\partial_t} \dfrac{\partial \Phi}{\partial s}, d\varphi_{s,t} \right\rangle + \left\langle R^N \left(d\varphi_{s,t}, \dfrac{\partial \Phi}{\partial t} \right) \dfrac{\partial \Phi}{\partial s}, d\varphi_{s,t} \right\rangle$$
$$+ \left\langle \nabla^{\Phi} \dfrac{\partial \Phi}{\partial s}, \nabla_{\partial_t} d\varphi_{s,t} \right\rangle .$$

Taking into account (3), we conclude

(5)
$$H_\varphi(v,w) = \int_M \left\langle d\nabla^{\Phi}_{\partial_t} \dfrac{\partial \Phi}{\partial s}, d\varphi_{s,t} \right\rangle \bigg|_{0,0} dx$$
$$- \int_M \text{Trace} < R^N(d\varphi, w)d\varphi, v > dx + \int_M < \nabla^{\varphi}_v, \nabla^{\varphi}_w > dx .$$

Harmonicity of φ insures that the first integral

$$\int_M \left\langle d\nabla^{\Phi}_{\partial_t} \dfrac{\partial \Phi}{\partial s}, d\varphi_{s,t} \right\rangle \bigg|_{0,0} dx = \int \left\langle \nabla^{\Phi}_{\partial_t} \dfrac{\partial \Phi}{\partial s}, d^*d\varphi_{s,t} \right\rangle \bigg|_{0,0} dx = 0 .$$

Thus we have

(6) **Proposition.** *The Hessian of a harmonic map φ is given by*

$$H_\varphi(v,w) = \int_M \left(< \nabla^{\varphi}_v, \nabla^{\varphi}_w > - < \text{Trace } R^N(d\varphi, v)d\varphi, w > \right) dx$$
$$= \int_M < \Delta^{\varphi} v - \text{Trace } R^N(d\varphi, v)d\varphi, w > dx ,$$

where $\Delta^{\varphi} = -\text{Trace } \nabla \nabla^{\varphi}$.

(7) $J_\varphi = \Delta^{\varphi} - \text{Trace } R^N(d\varphi, \cdot)d\varphi$ is the *Jacobi operator of E*. The variations in its kernel are called *Jacobi fields of φ*. Now J_φ is a self-adjoint linear elliptic differential operator. Therefore, the following numbers are finite:

nullity $\varphi = \dim \text{Ker } J_\varphi$ and

index $\varphi = E$-$\text{index}(\varphi) = $ dimension of the largest subspace of $C(\varphi^{-1}T(N))$ on which the quadratic form H_φ is negative definite.

A harmonic map φ of E-index 0 is said to be *E-stable*, or simply *stable*.

(8) **Example.** If the sectional curvature of N is non–positive, then we deduce immediately from Proposition (6) that the E–index of any harmonic map $\varphi : M \to N$ is zero; and so φ is stable.

(9) Next, suppose that $j : N \to \mathbf{R}^r$ is a Riemannian immersion. Take any vector $v \in \mathbf{R}^r$ and define the vector field $V \in C(T(N))$ at $y \in N$ by

$$V(y) = \nabla <j(y), v>.$$

That determines the special variation $\varphi_t(x) = \exp tV(\varphi(x))$ of a harmonic map $\varphi : M \to N$.

We define the quadratic form $Q_\varphi : \mathbf{R}^r \to \mathbf{R}$ by

(10) $$Q_\varphi(v) = \left.\frac{d^2 E(\exp tV \circ \varphi)}{dt^2}\right|_0.$$

In (11–19) below we shall perform computations by choosing orthonormal frames at given points; it is a routine exercise to verify that the relevant integrands are well defined functions.

Letting β denote the second fundamental form of N in $\mathbf{R}^r$, we have

(11) **Lemma.** *If $(e_i)_{1 \leq i \leq m}$ is an orthonormal base of $T_x(M)$ and $(v_\alpha)_{1 \leq \alpha \leq n}$ an orthonormal base of $T_{\varphi(x)}(N)$, then*

(12)
$$\text{Trace } Q_\varphi = \int_M \sum_{\alpha,i} \left[2\|\beta(d\varphi(e_i), v_\alpha)\|^2 - <\beta(d\varphi(e_i), d\varphi(e_i)), \beta(v_\alpha, v_\alpha)>\right] dx.$$

Proof. Step 1. By definition, $\beta(X, Y) = (\nabla_X^{\mathbf{R}^r} Y)^\perp$ for any $X, Y \in C(T(N))$. It is convenient to define the shape operator at $y \in N$ corresponding to a vector $Z \in T_y^\perp(N)$ by $B^Z(X) = -(\nabla_X^{\mathbf{R}^r} Z)^T$ for $X \in C(T(N))$, where superscript T denotes the tangential component.

Then B^Z is self–adjoint and satisfies

$$<B^Z(X), Y> = <\beta(X, Y), Z>.$$

For any parallel vector field V on $\mathbf{R}^r$ we have

$$\nabla_{d\varphi(e_i)} V^T = \left(\nabla_{d\varphi(e_i)}^{\mathbf{R}^r}(V - V^\perp)\right)^T = -\left(\nabla_{d\varphi(e_i)}^{\mathbf{R}^r} V^\perp\right)^T$$
$$= B^{V^\perp}(d\varphi(e_i)).$$

Step 2. At any point $\varphi(x) \in N$ let $(V_\lambda)_{1 \leq \lambda \leq r}$ be a local orthonormal frame in $\mathbf{R}^r$, with $(V_\alpha)_{1 \leq \alpha \leq n} = (v_\alpha)_{1 \leq \alpha \leq n}$ at $\varphi(x)$. Then for each i we have the pointwise identity

$$\sum_{\gamma=n+1}^{r} \|B^{V_\gamma}(d\varphi(e_i))\|^2 = \sum_{\alpha,\gamma} <B^{V_\gamma}(d\varphi(e_i)), V_\alpha>^2$$

$$= \sum_{\alpha,\gamma} <\beta(d\varphi(e_i), V_\alpha), V_\gamma>^2 = \sum_{\alpha=1}^{n} \|\beta(\varphi(e_i), V_\alpha)\|^2 .$$

Step 3. Treat each V_α as a parallel field as in Step 1:

$$\int_M \sum_{i,\alpha} \|\beta(d\varphi(e_i), V_\alpha)\|^2 dx = \int_M \sum_{i,\gamma} \|B^{V_\gamma}(d\varphi(e_i))\|^2 dx$$

$$= \int_M \sum_{i,\alpha} \|\nabla_{d\varphi(e_i)} V_\alpha\|^2 dx .$$

Step 4. Gauss' equation for N in $\mathbf{R}^r$ is

$$<R(X,Y)Z, W> = <\beta(X,Z), \beta(Y,W)> - <\beta(X,W), \beta(Y,Z)> .$$

Thus
$$<R(d\varphi(e_i), V_\alpha)d\varphi(e_i), V_\alpha> =$$
$$<\beta(d\varphi(e_i), d\varphi(e_i)), \beta(V_\alpha, V_\alpha)> - \|\beta(d\varphi(e_i), V_\alpha)\|^2 .$$

Step 5. From (6) we find

$$Q_\varphi(V_\alpha) = \int_M \sum_i \left(\|\nabla_{d\varphi(e_i)} V_\alpha\|^2 - <R(d\varphi(e_i), V_\alpha)d\varphi(e_i), V_\alpha> \right) dx .$$

Substituting from Steps 3 and 4 we obtain

$$\text{Trace } Q_\varphi = \int_M \sum_{i,\alpha} \left(2\|\beta(d\varphi(e_i), V_\alpha)\|^2 - <\beta(d\varphi(e_i), d\varphi(e_i)), \beta(V_\alpha, V_\alpha)> \right) dx . //$$

(13) Let $j : N \to \mathbf{R}^r$ be a Riemannian immersion, as in (9). A second application of (12) gives

$$\text{Trace } Q_\varphi = \int_M \sum_{i=1}^{m} \left[<\beta(d\varphi(e_i), d\varphi(e_i)), \tau(j)> -2 \text{Ricci}^N(d\varphi(e_i), d\varphi(e_i)) \right] dx ,$$

where $\text{Ricci}^N(X,Y) = \sum_{\alpha=1}^{n} < R^N(X,V_\alpha)V_\alpha, Y >$.

(14) Lemma. *If $j : N \to S^{r-1}$ is a compact minimally immersed submanifold of S^{r-1}, then*

$$< \beta(X,Y), \tau(j) > = n < X,Y > \quad \text{for all } X, Y \in \mathcal{C}(T(S^{r-1})).$$

Proof. Use Proposition I (1.7):

$$< \nabla dj(X,Y), \Delta(j) > = \ll X,Y > j, |dj|^2 j \gg = n < X,Y > . \; //$$

(15) Proposition. *Let $j : N \to S^{r-1}$ be a compact minimally immersed submanifold of S^{r-1}; and assume that $\text{Ricci}^N > n/2$. Then every E-stable harmonic map $\varphi : M \to N$ is constant.*

Proof. Let $\rho = \min \text{Ricci}^N$; and $\varphi : M \to N$ a non-constant harmonic map. Applying (14) to (13),

$$\text{Trace } Q_\varphi = 2n E(\varphi) - 2 \int_M \sum_i \text{Ricci}^N (d\varphi(e_i), d\varphi(e_i)) dx$$

$$\leq 2(n - 2\rho) E(\varphi) < 0,$$

by our hypotheses; i.e., φ is E-unstable. //

(16) Now return to the situation in (9), and let $\psi : N \to P$ be a harmonic map. For each $v \in \mathbf{R}^r$ we have the special variation ψ_t of ψ defined by $\psi_t(y) = \psi(\exp tV(y))$. Then

$$\left. \frac{\partial \psi_t}{\partial t} \right|_0 = d\psi \cdot V.$$

(17) Define the quadratic form $_\psi Q : \mathbf{R}^r \to \mathbf{R}$ by

$$_\psi Q(v) = \left. \frac{d^2 E(\psi \circ \exp tV)}{dt^2} \right|_0.$$

In terms of that special variation, calculation *ab initio* using Weitzenböck's formula yields

(18) **Lemma.**

$$\text{Trace }_\psi Q = \int_N \sum_{\alpha,\gamma,\lambda=1}^n [2 < \beta(v_\alpha, v_\lambda), \beta(v_\lambda, v_\gamma) >$$
$$- < \beta(v_\alpha, v_\gamma), \beta(v_\lambda, v_\lambda) >] < d\psi(v_\alpha), d\psi(v_\gamma) > dy\,.$$

Here the integrand is calculated in terms of an orthonormal frame (v_λ) in $T_y(N)$.

(19) **Proposition.** *With the assumptions of Proposition* (15) *every E–stable harmonic map* $\psi : N \to P$ *is constant.*

Proof. Suppose ψ is non–constant. In Lemma (18) choose (v_λ) at $y \in N$ so that $< d\psi(v_\alpha), d\psi(v_\gamma) > = a_\alpha\,\delta_{\alpha\gamma}$ with each $a_\alpha \geq 0$. As in the proof of (15),

$$\text{Trace }_\psi Q = \int_N \sum_{\alpha=1}^n [< \beta(v_\alpha, v_\alpha), \tau(j) > -2\,\text{Ricci}^N(v_\alpha, v_\alpha)]\,a_\alpha\,dy$$
$$\leq 2(n-2\rho)\,E(\psi) < 0\,.$$

Thus ψ is seen to be E–unstable. //

Refs. Mazet [Ma], Smith [Sm3], Ohnita [O], Howard–Wei [HW].

The identity map $Id_M : M \to M$.

(20) Using the metric on M to define the bundle isomorphism $\flat : T(M) \to T^*(M)$ and its inverse $\# : T^*(M) \to T(M)$, we can form the differential operators on vector fields $v \in C(T(M))$:

$$\bar{d}v = (dv^\flat)^\#, \quad \bar{d}^*v = d^*v^\flat = \text{div }v, \quad \bar{\Delta}v = (\Delta v^\flat)^\#\,;$$

here Δ denotes the de Rham–Hodge operator on 1–forms.

(21) **Lemma.** *Letting* L_v *denote the Lie derivative with respect to* $v \in C(T(M))$, *we have*

$$\int_M < Jv, v > dx = \int_M \left(\frac{1}{2}|L_v g|^2 - (\bar{d}^*v)^2\right) dx\,.$$

Proof. That is obtained by substituting the appropriate terms from (22) and (23) into (24):

$$L_v g(X,Y) = < \nabla_X v, Y > + < X, \nabla_Y v >, \text{ so}$$

(22)
$$\frac{1}{2}|L_v g|^2 = \text{Trace }(\nabla v \cdot \nabla v) + |\nabla v|^2\,.$$

$$\frac{1}{2}\Delta |v|^2 = < \Delta v, v > - |\nabla v|^2, \text{ where } \Delta = -\text{Trace }\nabla^2$$
$$= < J_{Id}v, v > + < \text{Ricci }v, v > - |\nabla v|^2\,;$$

so by the divergence theorem,

(23) $$0 = \int_M (<J_{Id}v, v> + <\text{Ricci } v, v> - |\nabla v|^2) \, dx .$$

$\text{div}(<v, \nabla v> -v \text{ div } v) = \text{Trace}(\nabla v \cdot \nabla v) + <\text{Ricci } v, v> -(\text{div } v)^2$, so

(24) $$0 = \int_M (\text{Trace}(\nabla v \cdot \nabla v) + <\text{Ricci } v, v> -(\text{div } v)^2) \, dx . \quad //$$

We compare the Jacobi operator $J_{Id}(v) = \Delta v - \text{Ricci}^M v$ to Weitzenböck's formula for the Laplacian on 1-forms, which can be written $\bar{\Delta} v = \Delta v + \text{Ricci}^M v$:

(25) $$\bar{\Delta} - J_{Id} = 2 \, \text{Ricci}^M .$$

(26) The *scalar curvature* of M is the function

$$\text{Scal}^M = \text{Trace Ricci}^M = \sum_{i,j=1}^m <R^M(e_i, e_j)e_i, e_j>,$$

relative to any orthonormal base (e_i) at any point.

Say that M is an *Einstein manifold* if its Ricci curvature satisfies $\text{Ricci}^M = cg$ for some $c \in \mathbf{R}$. It follows that $c = \text{Scal}^M/m$. And (25) becomes

(25') $$\bar{\Delta} - 2cI = J_{Id} .$$

(27) **Proposition.** *Suppose that M is a compact Einstein manifold. Id_M is E-unstable iff $\lambda_1 < 2 \, \text{Scal}^M/m$ where λ_1 is the first positive eigenvalue of Δ on functions.*

Proof. We shall show that Index (Id_M) is the number of eigenvalues λ of Δ such that $0 < \lambda < 2c$. Firstly, if (f, λ) is an eigenpair with $\lambda < 2c$, then $v = (df)^\#$ is a vector field satisfying $\bar{\Delta} v = \lambda v$; and $H_{Id}(v, v) < 0$, by (25') and Proposition (6).

Conversely, if $v \in C(T(M))$ satisfies $\bar{\Delta} v = \lambda v$ with $\lambda < 2c$, then its Hodge decomposition $v^\flat = df + \sigma$ with $d^*\sigma = 0$. Then $\Delta df = \lambda df$ and $\Delta \sigma = \lambda \sigma$. We claim that $\sigma \equiv 0$; for otherwise, $\int_M <J_{Id} \sigma^\#, \sigma^\#> dx < 0$; but by Lemma (21) we have

$$\int_M <J_{Id} \sigma^\#, \sigma^\#> dx = \frac{1}{2} \int |L_{\sigma^\#} g|^2 \, dx \geq 0 ,$$

which is a contradiction.

We conclude that $v = (df)^\#$; and by adding a suitable constant, we have $\Delta f = \lambda f$. In conclusion, the space on which H_{Id} is negative definite is generated by the differentials of eigenfunctions of Δ with positive eigenvalues $< 2c$. //

(28) **Example.**
$$\text{Index } (Id_{S^m}) = 0 \text{ for } m = 1 \text{ or } 2$$
$$= m + 1 \text{ for } m \geq 3.$$

For $m = 1$ or 2, Id_{S^m} is an absolute minimum of E, by elementary considerations. For $m \geq 3$, $c = m - 1$ and $\lambda_1 = m$ by VIII (1.9). And that is the only eigenvalue λ such that $0 < \lambda < 2(m-1)$. The mutliplicity of λ_1 is $m + 1$.

(29) **Theorem** (Howard–Wei, Ohnita). *Let N be a compact n–dimensional irreducible homogeneous space. The following properties are equivalent:*

a) $\lambda_1 < 2\text{Scal}^N/n$, *where λ_1 is the smallest positive eigenvalue of Δ on functions.*

b) *The identity map Id_N is E-unstable.*

c) *Every E-stable harmonic map $\varphi : M \to N$ from a compact manifold M is constant.*

d) *Every E-stable harmonic map $\psi : N \to P$ is constant.*

Proof. a) $\Leftrightarrow$ b) is the content of Proposition (27), because N is an Einstein manifold. c) $\Rightarrow$ b) and d) $\Rightarrow$ b) are both obvious.

Now assume a); then $\lambda_1 < 1$, because the metric g_0 on N has scalar curvature $= n/2$. Let N_1 denote N with the metric $g_1 = \dfrac{\lambda_1}{n} g_0$. Then $\text{Scal}^{N_1} = n^2/2\lambda_1$. And $\text{Ricci}^{N_1} \equiv \rho_1 = n/2\lambda_1$; so $n - 2\rho_1 = n(1 - 1/\lambda_1) < 0$. We can now apply Propositions (15) and (19) to conclude a) $\Rightarrow$ c) and a) $\Rightarrow$ d). //

(30) **Example.** Take $N = S^m$ with $m \geq 3$. Then by (28) we see that Id_{S^m} is unstable. *Therefore, every stable harmonic map $\psi : S^m \to N$ is constant. And every stable harmonic map $\varphi : M \to S^m$ is constant, provided M is compact.*

If $\psi : S^m \to N$ has maximum rank $k \geq 1$, then E–index $(\psi) \geq k + 1$. Supposing that M is compact, a non–constant harmonic map $\varphi : M \to S^n$ has E-index $(\varphi) \geq m - 2$; equality holds iff $\varphi(M)$ is contained in a totally geodesic 2-sphere.

(31) Remark. It is known that the conditions in Theorem (29) insure that the homotopy groups $\pi_1(N) = 0 = \pi_2(N)$. Thus in contrast to Example (30), if $\Gamma \neq \{e\}$ is a finite group of isometries acting freely on S^m, then $Id_{S^m/\Gamma}$ is stable. In particular, the identity map on real projective m–space is stable.

(32) Proposition. *Let $\varphi : M \to S^n$ be a non–constant harmonic map of a compact manifold, and let $j : S^n \to S^{n+r}$ be the standard totally geodesic inclusion. Set $\Phi = j \circ \varphi$.*

Then

$$\text{(33)} \qquad E\text{–index } \Phi \geq E\text{–index } \varphi + r \sum_{k=0}^{K} m_k ,$$

where $K = \max\{k : \lambda_k \leq 2e(\varphi) \text{ but } \lambda_k \not\equiv 2e(\varphi)\}$ and m_k = multiplicity of λ_k. If φ is an eigenmap (i.e., $e(\varphi)$ is constant), then we have equality in (33).

Proof. Let W denote the space of variations w of Φ normal to $\varphi(M)$ for which $H_\Phi(w, w) < 0$.

For (33) it is sufficient to verify that

$$\dim W \geq r \sum_{k=0}^{K} m_k .$$

For each k let $\{f_j^k : 1 \leq j \leq m_k\}$ be an orthonormal base for the eigenspace $E(\lambda_k)$. And $\{u_\alpha : 1 \leq \alpha \leq r\}$ an orthonormal frame field normal to S^n in S^{n+r}, with each u_α parallel; we treat these as variations of Φ. Because the curvature tensor of a sphere satisfies

$$R(X,Y)Z = <X, Z> Y - <Y, Z> X ,$$

the Jacobi operator

$$J_\Phi(f_j^k u_\alpha) = \Delta(f_j^k u_\alpha) - |d\varphi|^2 f_j^k u_\alpha$$
$$= (\lambda_k - |d\varphi|^2) f_j^k u_\alpha .$$

Thus

$$H_\Phi(f_j^k u_\alpha, f_j^k u_\alpha) = \int_M (\lambda_k - |d\varphi|^2)|f_j^k|^2 dx < 0 .$$

Consequently, the variations

$$\{f_j^k u_\alpha : 0 \leq k \leq K, 1 \leq j \leq m_k, 1 \leq \alpha \leq r\}$$

are independent elements of W.

Now assume that $e(\varphi)$ is constant. To show equality in (33) it is sufficient to establish (34) and (35) below.

(34) *If w is a variation of Φ such that $J_\Phi(w) = -\lambda w$ for some $\lambda > 0$, then $J_\Phi(w^\perp) = -\lambda w^\perp$.* Indeed, $J_\Phi w^T = -\lambda w - J_\Phi w^\perp$. Taking the normal components of both sides gives $0 = -\lambda w^\perp - (J_\Phi w^\perp)^\perp$. But $J_\Phi w^\perp$ is in fact normal to S^n.

(35) $$\dim W = r \sum_{k=0}^{K} m_k .$$

Because of the independence of $\{u_\alpha : 1 \leq \alpha \leq r\}$, it is sufficient to prove (35) in case $r = 1$. Set $u = u_1$; then if $w \in W$, we have $w = fu$ for some function $f : M \to \mathbf{R}$. As above,

$$J_\Phi(w) = \Delta f \cdot u - |d\varphi|^2 fu .$$

Writing $J_\Phi(w) = -\lambda w$ for some $\lambda > 0$, we see that

(36) $$\Delta f = (-\lambda + |d\varphi|^2) f .$$

Thus the constant $-\lambda + |d\varphi|^2 = \lambda_k$ for some k; and $0 < \lambda = |d\varphi|^2 - \lambda_k$ iff $1 \leq k \leq K$. Consequently, (36) has $\sum_{k=0}^{K} m_k$ linearly independent solutions; i.e.,

$$\dim W = \sum_{k=0}^{K} m_k . //$$

(37) **Example.** Let $j : S^m \to S^n$ be a standard totally geodesic embedding $(m < n)$. By Proposition (32), E–index $(j) = E$–index $(Id_{S^m}) + n - m$. From (28) we find
$$E\text{–index } (j) = \begin{cases} n+1 & \text{if } m \geq 3 \\ n-2 & \text{if } m = 2 . \end{cases}$$

(38) Here is an example to show how the choice of metrics can influence the E–index of a harmonic map. Let $g = g_1$ be the standard metric on the sphere $S^3 = S^3(1)$ of radius 1; and $\varphi : S^3 \to S^2(1/2)$ the Hopf fibration of VIII (2.8), which is Riemannian. Let T^V denote the subbundle of $T(S^3)$ consisting of vectors

along the fibres of φ; and T^H the bundle of vectors in $T(S^3)$ orthogonal to the fibres. For any $t > 0$ set

$$g_t = \begin{cases} g & \text{on} \quad T^H \times T^H \\ 0 & \text{on} \quad T^H \times T^V \text{ and } T^V \times T^H \\ t^2 g & \text{on} \quad T^V \times T^V \end{cases}.$$

If η is the vertical unit Killing field on S^3, then $g_t = g + (t^2 - 1)\eta^b \odot \eta^b$. These (g_t) are the *Berger metrics* on S^3. For each t, $\varphi : (S^3, g_t) \to S^2(1/2)$ is a Riemannian fibration with fibres which are closed geodesics. These are all harmonic Riemannian submersions.

(39) Urakawa [U2] has shown that *E–index* $(\varphi) = 4$ *for* $t = 1$; *and that there is* $\varepsilon > 0$ *such that E–index* $(\varphi) = 0$ *for* $0 < t < \varepsilon$.

A similar phenomenon had been observed by Smith [Sm1, 3] for the projection map $S^m \times S^n \to S^n$ ($n \geq 3$).

(40) **Remark.** Special variations of maps to and from spheres S^n are obtained from conformal vector fields on S^n. That method was exploited by Simons [Si] – and is widely used throughout this Appendix (e.g., in proofs of the assertions in (30)).

Refs. El Soufi [ElS], Howard–Wei [HW], Leung [Leu], Mazet [Ma], Ohnita [O], Pluzhnikov [Pl2], Sealey [Se], Smith [Sm3], Tyrin [Ty], Urakawa [2], Xin [X1]. In Theorem (29) the equivalence b) $\Leftrightarrow$ d) was found by Pluzhnikov; and b) $\Leftrightarrow$ c) by Tyrin.

Ohnita [O] and Urakawa [U1] have classified the compact irreducible symmetric spaces with unstable identity maps.

V–index of a minimal immersion

(41) For simplicity, suppose that M is compact. Take a 2–parameter variation $\varphi_{s,t}$ by immersions of a minimal immersion $\varphi_{0,0} = \varphi : M \to N$, writing

$$v = \left.\frac{\partial \varphi_{s,t}}{\partial s}\right|_{0,0} \qquad w = \left.\frac{\partial \varphi_{s,t}}{\partial t}\right|_{0,0}.$$

The V–Hessian of φ is

(42) $$H^V_\varphi(v,w) = \left.\frac{\partial^2 V(\varphi_{s,t})}{\partial t \partial s}\right|_{0,0}.$$

In analogy with Proposition (6) we have

(43) **Proposition.**

$$H_\varphi^V(v, w) = \int_M < J_\varphi^V(v^\perp), w^\perp > dx \,,$$

where as in (7)

(44)
$$-J_\varphi^V(v^\perp) = \text{Trace } (\nabla^\perp)^2 v^\perp + \text{Trace } \beta(\varphi)(A^{v^\perp})$$
$$+ \text{Trace } R^N(d\varphi, v^\perp)(d\varphi)^\perp$$

and $A^{v^\perp} \in C(T^*(M) \otimes T(N))$ is characterized by

$$d\varphi(A^{v^\perp}(u)) = -(\nabla_u v^\perp)^T \quad \text{for all} \quad u \in C(T(M))\,.$$

Again, J_φ^V is a self–adjoint linear elliptic differential operator. We define V–nullity(φ), V–index(φ), and V–stability in analogy with (7).

(45) *If v, w are normal variations of φ, then*

$$H_\varphi^E(v, w) - H_\varphi^V(v, w) = 2 \int_M < (\nabla v)^T, (\nabla w)^T > dx \,.$$

(46) Simons [Si] has shown that *if $\varphi : M \to S^n$ is a minimal immersion, then*

$$V\text{–index}(\varphi) \geq n - m$$

with equality iff M is immersed onto a totally geodesic sphere. In that latter case, V–index $(\varphi) = E$–index $(\varphi) = n - 2$ if $m = 2$. By way of contrast, for the standard embedding $j : S^m \to S^n$ with $m \geq 3$, V–index $(\varphi) = n - m < n + 1 = E$–index (φ), by (37).

El Soufi [ElS] has strengthened Simons' theorem by showing that *if the image of a minimal immersion $\varphi : M \to S^n$ is not a totally geodesic sphere, then V–index $(\varphi) \geq n + 1$.*

Refs. Simons [Si], El Soufi [ElS].

APPENDIX 2.   RIEMANNIAN IMMERSIONS $S^m \to S^n$

(1) Let $\varphi : M \to N$ be a Riemannian immersion. For $c \in \mathbf{R}$ and $x \in M$ let

$$T_x^{(c)} = \{X \in T_x(M) : R^M(X,Y)Z$$
$$= c(<X,Z>Y - <Y,Z>X) \text{ for all } Y, Z \in T_x(M)\}.$$

Set

$\mu_x^{(c)} = \dim T_x^{(c)}$, the *c–nullity of M at x*. Define

$K_x = \{X \in T_x(M) : \beta_\varphi(X,Y) = 0 \text{ for all } Y \in T_x(M)\}$, and

$\nu_x = \dim K_x$, the *relative nullity of φ at x*.

(2) **Proposition.** *Let $\varphi : M \to N(c)$ be a Riemannian immersion in a space form. Then for all $x \in M$*

(3) $$\nu_x \leq \mu_x^{(c)} \leq \nu_x + n - m$$

where $m = \dim M$ and $n = \dim N(c)$.

Those inequalities are due to Chern–Kuiper (see [KNII, Note 16]; the proof there is given for $c = 0$, but it applies as well to any real number c).

(4) We consider Riemannian immersions $\varphi : S^m \to S^n$ of Euclidean spheres of unit radii. Then $\mu_x^{(1)} = m$, and if $n < 2m$ we have $0 < 2m - n \leq \nu_x$ for all $x \in S^m$. Set

$$\nu = \min\{\nu_x : x \in S^m\} \quad \text{and} \quad U = \{x \in S^m, \nu_x = \nu\}.$$

Then ν is upper semi–continuous and U is open (because every $x_0 \in S^m$ has a neighbourhood U_0 on which $\nu_x \leq \nu_{x_0}$ for all $x \in U_0$).

Now $K = \cup\{K_x : x \in U\}$ is a ν–dimensional subbundle of $T(S^m)$. Take $x \in U$ and a section X of K with $X_x \neq 0$; and Y, Z sections of $K^\perp$. Because $\beta_\varphi(K_y, T_y(S^m)) = 0$ for all $y \in U$, we apply Codazzi's equation (setting $\beta = \beta_\varphi$):

$$(\nabla_X \beta)(Y,Z) = (\nabla_Y \beta)(X,Z) = \nabla_Y(\beta(X,Z)) - \beta(\nabla_Y X, Z) - \beta(X, \nabla_Y Z)$$
$$= \beta(-\nabla_Y X, Z) = \beta(C_X Y, Z),$$

where $C_X Y = \mathrm{Proj}_{K^\perp}(-\nabla_Y X)$.

The symmetry of $\nabla_X \beta$ implies that C_X is self-adjoint; i.e.,

$$\beta(C_X Y, Z) = \beta(Y, C_X Z).$$

(5) Let $\gamma = \gamma_\varphi$ denote the Gauss map of φ; thus $\beta_\varphi = d\gamma_\varphi$. The *third fundamental form* of φ is $\gamma_\varphi^* k$, where k is the canonical Riemannian metric of the Grassmannian; and

$$(\gamma^* k)(Y, Z) = \sum_i < \beta(Y, X_i), \beta(Z, X_i) > = \sum_i < \beta(Y, Z), \beta(X_i, X_i) >$$

by Gauss' equation, where (X_i) is a local orthonormal frame. From that we find

(6) $$(\gamma_\varphi^* k)(C_X Y, Z) = (\gamma_\varphi^* k)(Y, C_X Z).$$

(7) The bundle $K \to U$ is integrable, and its leaves are totally geodesic submanifolds [KNII, Note 16]; thus if X, W are sections of K, then so is $\nabla_W X$.

(8) **Lemma.** *Let $\alpha : \mathbb{R} \to S^m$ be a geodesic with $\alpha'_x \in K_x$ and $|\alpha_x|^2 \equiv 1$. Letting X be a section of K extending α' with $|X_x|^2 \equiv 1$, we have*

(9) $$\nabla_X^\perp C_X = C_X^2 + Id \circ \alpha,$$

operating on K.

Proof. Because K is totally geodesic, for any sections W of K and V of $T(M)$,

(10) $$P(\nabla_W V) = P(\nabla_W P(V)) = \nabla_W P(V)$$

where $P : T(M) \to K^\perp$ denotes orthogonal projection. Then for a section Y of $K^\perp$,

(11) $$\begin{aligned}(\nabla_X^\perp C_X)Y &= -\nabla_X^\perp P(\nabla_Y X) + P(\nabla_{\nabla_X^\perp Y} X) \\ &= -P(\nabla_X \nabla_Y X - \nabla_{\nabla_X Y} X) \\ &= P(R^{S^m}(X, Y)X - \nabla_Y \nabla_X X + \nabla_{\nabla_X Y} X).\end{aligned}$$

(12) But $P(R^{S^m}(X, Y)X) = Y$. And $P(\nabla_Y \nabla_X X) = 0$ along α; for if Z is a section of $K^\perp$,

$$< \nabla_Y \nabla_X X, Z > = Y < \nabla_X X, Z > - < \nabla_X X, \nabla_Y Z >;$$

the first term in the right hand member vanishes because K is totally geodesic; the second vanishes along α by definition of X.

Now (10) insures that $P(\nabla_{\nabla_Y X} X) = C_X^2 Y$. Together with (11) and (12), that shows that (9) is satisfied along α. //

(13) **Lemma.** $C_X | K_x^\perp$ *has no real eigenvalues for any $x \in U$, provided $X \neq 0$.*

Proof. Because C_X satisfies (9) at every $\alpha(t)$, we are reduced to studying the equation

$$\dot{A} + A^2 + I = 0$$

for a solution $A : \mathbb{R} \to \text{End } V$, where V is a finite dimensional vector space. If $(\lambda(t), X(t))$ is an eigenpair of $A(t)$, then

(a) $\dot{\lambda} + \lambda^2 + 1 = 0$; and

(b) the $\lambda(t)$-eigenspace $V_t = V_0$. Thus for any eigenpair $(\lambda(0), X(0))$ and any $t \in \mathbb{R}$,

$$A(t) X(0) = \lambda(t) X(0).$$

Because $\lambda(t) = \dfrac{\lambda(0) - \tan t}{\lambda(0) \tan t + 1}$ and $\lambda(t) \in \mathbb{C}$ is defined for all $t \in \mathbb{R}$, we conclude that
Im $\lambda(0) \neq 0$; and from that, Im $\lambda(t) \neq 0$ for all $t \in \mathbb{R}$. //

The following result – and stronger forms of it – are due to Moore [Moo]. We follow the treatment of Ferus [F].

(14) **Theorem.** *If $\varphi : S^m \to S^n$ is a Riemannian immersion and $n < 2m$, then φ is totally geodesic.*

Proof. To show $\beta_\varphi \equiv 0$ it suffices to show $K^\perp = 0$. But the quadratic form $\gamma_\varphi^* k$ is positive definite on $K^\perp$, and with respect to it, C_X is self-adjoint. By Lemma (13) that is impossible unless $K^\perp = 0$. //

(15) **Example.** The following composition gives a Riemannian immersion of $S^m(1)$ in $S^{2m+1}(1)$, which is certainly not totally geodesic:

$$\begin{array}{c} S^m \hookrightarrow \mathbb{R}^{m+1} \\ \downarrow \\ T^{m+1} \hookrightarrow S^{2m+1} \end{array}.$$

Here
$$T^{m+1} = \mathbf{R}^{m+1}/(2\pi/\sqrt{m+1})\, Z^{m+1}$$
$$= S^1(1/\sqrt{m+1}) \times \ldots \times S^1(1/\sqrt{m+1}),$$

which is naturally embedded isometrically (with curvature zero) in the unit sphere of $\mathbb{C}^{m+1}$.

(16) **Example** [FP]. *There are Riemannian embeddings of $S^2(1)$ in $S^4(1)$ which are not totally geodesic.*

The idea is to treat $S^4 = \{A \in S \text{ End } \mathbf{R}^3 : \text{Trace } A = 0 \text{ and } |A|^2 = 1\}$. Here S End $\mathbf{R}^3$ is the vector space of self–adjoint endomorphisms of $\mathbf{R}^3$ with inner product $k = \frac{1}{2}$ Trace AB. The rotation group $SO(3)$ acts on S^4 by conjugation; its regular orbits form an isoparametric family of hypersurfaces (IV (3.5)). The orbit M_0 of $\begin{pmatrix} 0 & & \\ & 1 & \\ & & -1 \end{pmatrix}$ is the unique minimal orbit. $S^3 = \text{Spin}(3) \xrightarrow{\pi} M_0$ is an 8–fold cover. We endow S^3 with the metric $\pi^* k$. Then find Riemannian immersions $S^2 \to (S^3, \pi^* k) \subset S^4$; a certain range of these are in fact embeddings.

APPENDIX 3. MINIMAL GRAPHS AND PENDENT DROPS

(1) The *graph* of a function $\varphi : \mathbf{R}^m \to \mathbf{R}$ is the map $\Phi : \mathbf{R}^m \to \mathbf{R}^m \times \mathbf{R}^1 = \mathbf{R}^{m+1}$ defined by $\Phi(x) = (x, \varphi(x))$. It induces the metric k on $\mathbf{R}^m$ with components

$$k_{ij} = \delta_{ij} + \varphi_{x^i} \varphi_{x^j} \quad (1 \leq i, j \leq m).$$

An easy computation shows that Φ is *minimal* iff

(2) $$\operatorname{div} \frac{\nabla \varphi}{(1 + |\nabla \varphi|^2)^{1/2}} = 0.$$

Bernstein's problem refers to the nature of the solutions of (2). We refer to the book of Giusti [Gi] for a beautiful and thorough exposition – and in particular, for the role played by geometric measure theory.

Clearly every affine function $\varphi : \mathbf{R}^m \to \mathbf{R}$ (i.e., polynomial of first degree) is a solution. It is known that *for* $m \leq 7$ *every solution is affine; and that for* $m \geq 8$, *there are others*. The first such examples were constructed by Bombieri, de Giorgi, and Giusti [BDG] – in the context of equivariant reduction theory.

(3) Indeed, let $G = SO(p) \times SO(q)$ act on $\mathbf{R}^p \times \mathbf{R}^q \times \mathbf{R}$ in the standard manner on the first two factors; and trivially on the third. The orbit space

$$Q = \mathbf{R}^p \times \mathbf{R}^q \times \mathbf{R}/G = \{(u, v, z) \in \mathbf{R}^3 : u, v \geq 0\}$$

with its flat metric $h^Q = du^2 + dv^2 + dz^2$. The singular set is $\{(u, v, z) \in Q : uv = 0\}$. The projection map $\sigma : \mathbf{R}^{p+q+1} \to Q$ is given by $\sigma(x, y, z) = (|x|, |y|, z)$. And the volume function is (up to a constant factor)

$$V(u, v) = u^{p-1} v^{q-1}.$$

We look for minimal embeddings of $\mathbf{R}^{p+q}$ in $\mathbf{R}^{p+q+1}$ which are $SO(p) \times SO(q)$-invariant – and especially for those which are graphs of functions $\varphi : \mathbf{R}^{p+q} \to \mathbf{R}$ expressible in the form $\varphi(x, y) = \psi(u, v)$, where $u = |x|, v = |y|$. Then (2) takes the form

(4) $$(1 + \psi_v^2)\psi_{uu} - 2\psi_u \psi_v \psi_{uv} + (1 + \psi_u^2)\psi_{vv}$$
$$+ \left((p-1) \frac{\psi_u}{u} + (q-1) \frac{\psi_v}{v} \right) (1 + \psi_u^2 + \psi_v^2) = 0.$$

If $p, q \geq 4$, existence of non-trivial solutions of (4) defined on the whole quadrant $u, v \geq 0$ can be established by exhibiting explicit sub– and super–solutions. In the terminology of Chapter IV (3), such solutions provide cohomogeneity 2 minimal embeddings of $\mathbf{R}^{p+q}$ in $\mathbf{R}^{p+q+1}$.

(5) The shape of the surface of a pendent liquid drop is determined by the condition that its mean curvature be proportional to the distance below a horizontal reference plane. That leads to the capillarity equation

$$\text{div}\,\frac{\nabla\varphi}{(1+|\nabla\varphi|^2)^{1/2}} = \kappa\varphi \tag{6}$$

for a function $\varphi : \mathbb{R}^m \to \mathbb{R}$ and a constant $\kappa < 0$ (the sign indicating that the gravitational field points away from the reference plane – as in an upside–down capillary tube). Here $\varphi(x)$ measures the distance from the reference plane to the point x on the surface.

(7) Concus and Finn [CF2,1] and Bidaut–Veron [BV] have studied $SO(m)$–invariant solutions of (6). We summarize their analysis, which has features similar to those of Chapter VII.

Set $\varphi(|x|) = u(r)$ and $\kappa = -(m-1)$; then (6) becomes

$$\left(\frac{r^{m-1}u'}{\sqrt{1+u'^2}}\right)' = -(m-1)r^{m-1}u. \tag{8}$$

In general, solutions of (8) are not defined for all $r \geq 0$, so it is convenient to consider its parametric form: In the notation of Chapter VII we set $(u(s), r(s)) = (x(s), y(s))$ where s denotes arc length. Then (8) takes the form

$$\dot{x}\ddot{y} - \dot{y}\ddot{x} - (m-1)\frac{\dot{x}}{y} = -(m-1)x \tag{9}$$

on the orbit space $\mathbb{R}^{m+1}/SO(m) = \{(x,y) \in \mathbb{R}^2 : y \geq 0\}$. Note that we have the trivial solution $x \equiv 0$; and that if $(x(s), y(s))$ is a solution, then so is $(-x(s), y(s))$.

(10) The methods of Chapter VI (2) provide a complete existence and uniqueness theory for solutions γ starting at the singular boundary; i.e., $\gamma_{x_0}(0) = (x_0, 0)$. Indeed, for each $x_0 \in \mathbb{R}$ there is a unique solution $\gamma_{x_0}(s)$ which starts orthogonally to the x–axis, is defined for all $s \geq 0$, and spreads out indefinitely away from the x–axis, without limit sets or double points. For large s, all such solutions can be expressed in the form $\gamma = x(y)$ and (except for $x \equiv 0$) have oscillatory behaviour.

(11) As x_0 varies between 0 and $-\infty$, we observe the following qualitative behaviour of solutions (see also Figure 12 below):

i) If $-2 \leq x_0 < 0$, then $\gamma_{x_0}(s)$ projects simply on the y–axis; thus it determines a global solution of (8).

ii) If $x_0 \leq -2\sqrt{2}$, then tangents parallel to the x–axis appear. The corresponding solutions of (8) blow up in finite time – but they can be continued indefinitely as solutions of (9). Computer studies have shown $\bar{x}_0 = -2.5678$ as the separating value between cases i) and ii).

iii) If $x_0 \ll -2\sqrt{2}$, then $\gamma_{x_0}(s)$ approximates Delaunay's arcs of elliptic type for small $s > 0$. And arcs of hyperbolic type (with the characteristic double points) for small $s < 0$.

iv) As $x_0 \to -\infty$, numerical experiments suggest that the solution curves γ_{x_0} tend to a solution $\gamma_\infty = x(y)$, defined for all $y > 0$. γ_∞ provides an example of a singular solution $U(r)$ of (8) – and therefore of (6) – with a non–removable isolated singularity.

It is known that $U(r) \simeq -\frac{1}{r}$ for $r \simeq 0$; and that $U(r)$ is the unique singular solution of (8) which is concave near $r = 0$. It is conjectured that $U(r)$ is the only singular solution of (8).

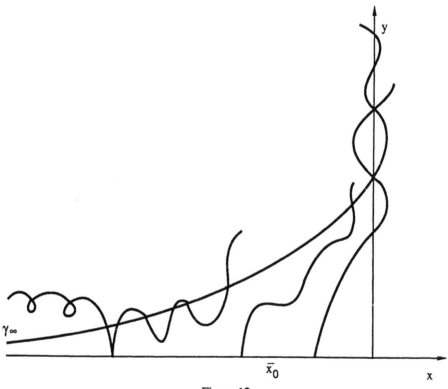

Figure 12

(12) Remark. The proof of local existence and uniqueness of solutions starting at the singular boundary given by Concus–Finn [CF1,2] was based on Schauder's fixed point theorem. Incidentally, they call attention to the possibility of applying the method of power series expansion, as in Chapter VI(2).

APPENDIX 4. FURTHER ASPECTS OF PENDULUM TYPE EQUATIONS

We begin with a proof of the comparison Lemma X (1.15). Then we illustrate a method, in the context of O.D.E., to establish the existence of harmonic joins between Euclidean spheres (Theorem IX (4.1) in the case $c^2 = 1 = d^2$).

For the proof of X (1.15) we shall need the following standard comparison theorem ([CL, p.210]);

(1) **Lemma.** *Let L_i ($i = 1, 2$) denote the differential operators*

$$L_i f = (p_i f')' + g_i f,$$

where $p'_i, g_i : [s, T] \to \mathbb{R}$ are continuous functions such that

(2) $$0 < p_2(t) \leq p_1(t) \quad \text{for all} \quad t \in [s, T]$$

(3) $$g_2(t) > g_1(t) \quad \text{for all} \quad t \in (s, T].$$

Let f_i be solutions of $L_i f_i = 0$ ($i = 1, 2$) on $[s, T]$, and define $w_i = \tan^{-1}(f_i / p_i f'_i)$. If

(4) $$w_2(s) \geq w_1(s),$$

then

(5) $$w_2(t) > w_1(t) \quad \text{for all} \quad (s, T].$$

For our application we choose

$$p_1(t) = \exp\left(\int_s^t D(u)du\right) = p_2(t);$$

$$g_1(t) = \tilde{G}(t) \frac{\sin F(t) \cos F(t)}{F(t)} p_1(t);$$

$$g_2(t) = G(t) \frac{\sin \bar{\beta}(t) \cos \bar{\beta}(t)}{\bar{\beta}(t)} p_2(t),$$

where G, D are defined in X (1.8). $\tilde{G}$ is defined through the equation

$$F''(t) + D(t)F'(t) + \tilde{G}(t) \sin F(t) \cos F(t) = 0;$$

and $\tilde{\beta}$ is the solution of X (1.8) described below.

We suppose

(6) $$A^+(s) < F'(s),$$

and derive a contradiction.

Indeed, (6) implies the existence of a solution $\tilde{\beta}$ of X (1.5) such that

(7) $$\tilde{\beta}(s) = 0, \quad \tilde{\beta}'(s) < F'(s), \quad \tilde{\beta}(T_1) = \pi/2$$

for some $T_1 > s$.

Let $T > s$ be the first time for which $\tilde{\beta}(T) = F(T)$; note that T is well defined because of assumption ii) in X (1.15). Now we can apply Lemma (1) on $[s, T]$, with $f_1 = F$ and $f_2 = \tilde{\beta}$. Indeed, hypotheses (2) and (4) are immediate, and (3) is a consequence of the inequalities

$$\tilde{\beta}(t) < F(t) \quad \text{and} \quad \tilde{G}(t) < G(t) \quad \text{on} \quad (s, T].$$

But $\tilde{\beta}'(T) \geq F'(T)$, and so $w_2(T) \leq w_1(T)$ – which contradicts (5). //

(8) Now we verify the assertion X (1.11). First of all we establish

(9) **Lemma.** *Suppose that* $A^+(s) > 0$. *Then*
i) *If* $b > A^+(s)$, *then* $\beta(s, b)$ *reaches* $\pi/2$ *in finite time*.
ii) *If* $0 < b < A^+(s)$, *then* $\beta(s, b)$ *turns down before reaching* $\pi/2$ *and thus vanishes at some* $T_1 > s$.

Proof. i) is obvious from the definition of $A^+(s)$. As for ii), it suffices to show that $\beta(s, b)$ cannot increase asymptotically to $\pi/2$. Indeed, set

$$f_1 = \beta(s, A^+(s)), \quad f_2 = \beta(s, b), \quad f = f_1 - f_2.$$

Applying Lemma (1) as above we see that $f(t) > 0$ for all $t > s$. If $f_2(t)$ increases asymptotically to $\pi/2$, then

(10) $$\int_s^{+\infty} f'(t) = f(+\infty) - f(s) = 0 \quad \text{with} \quad f'(s) = A^+(s) - b > 0.$$

Therefore there exists $T > s$ such that

$$f_1'(T) < f_2'(T), \quad f_1(T) > f_2(T).$$

We can again apply Lemma (1) on $[s, T]$ and see that (11) contradicts (5), precisely as in the proof of X (1.15) above. //

Solutions to X (1.8) with initial data

(12) $$\beta(s) = 0, \quad \beta'(s) = \tilde{b}$$

depend continuously on s and $\tilde{b}$. We express that as follows. Consider the initial value problem

(13) $$\begin{cases} \beta(s+\epsilon_1) = 0, \quad \beta'(s+\epsilon_1) = \tilde{b}+\epsilon_2 \quad \epsilon_1, \epsilon_2 \in \mathbb{R} \\ \beta(t) \text{ solves X (1.8)}. \end{cases}$$

(14)   As $\epsilon_1, \epsilon_2 \to 0$, solutions to (13) approximate on compact sets those of X (1.8), (12).

Now we are in position to prove X (1.11). We verify continuity at a point s for which $A^+(s) > 0$ (the case $A^+(s) = 0$ is similar). Suppose that $A^+(s)$ is not continuous at s; then there exist $K, \delta > 0$, $\epsilon_n \in \mathbb{R}$ such that $\epsilon_n \to 0$ as $n \to +\infty$ and either

(15) $$K \geq A^+(s+\epsilon_n) \geq A^+(s) + \delta \quad \text{for all } n$$

or

(16) $$A^+(s+\epsilon_n) \leq A^+(s) - \delta \quad \text{for all } n.$$

First we deal with case (15). Set $f_n = \beta(s+\epsilon_n, A^+(s+\epsilon_n))$: the f_n and their derivatives are uniformly bounded, and so they C^2-subconverge on compact sets to a solution $\bar{f}$ of X (1.8), by (14). The solution $\bar{f}$ satisfies

(17) $$\begin{cases} 0 \leq \bar{f}(t) \leq \pi/2 \quad \text{for all } t \geq s \\ \bar{f}'(s) \geq A^+(s) + \delta \,. \end{cases}$$

Therefore we conclude that $\bar{f}(t)$ increases asymptotically to $\pi/2$ as t increases from s to $+\infty$, so contradicting (9i).

Case (16) is similar: if $0 < \bar{f}'(s) \leq A^+(s) - \delta$, then the contradiction is obtained by using (9ii). If $\bar{f} \equiv 0$, then there would be solutions of X (1.8), (12) with $\tilde{b}$ arbitrarily small, which reach $\pi/2$ in finite time, contradicting the definition of $A^+(s)$. //

(18)   **Remark.** An approximation technique similar to (14) was used in the proof of VII (2.11).

(19) We now turn our attention to the equations arising in Theorem IX (4.1).

After substituting $\tan s = e^t (t \in \mathbf{R})$, the harmonicity equation IX (1.11) becomes

(20) $\qquad \beta''(t) + D(t)\beta'(t) + G(t) \sin \beta(t) \cos \beta(t) = 0$,

where
$$D(t) = \frac{(p-2)e^{-t} - (r-2)e^t}{e^t + e^{-t}}, \qquad G(t) = \frac{\lambda_v e^t - \lambda_u e^{-t}}{e^t + e^{-t}}.$$

The boundary conditions for $\beta : \mathbf{R} \to (0, \pi/2)$ are now

(21) $\qquad \lim_{t \to +\infty} \beta(t) = \pi/2, \quad \lim_{t \to -\infty} \beta(t) = 0$.

By interchanging the roles of the eigenmaps u, v if necessary, we can assume that $(DC)_2$ holds. Next, we write

(22) $\qquad \beta(t) = \tan^{-1} \left[\exp \left(\int_0^t H(s)ds + c \right) \right]$

for some $c \in \mathbf{R}$; the function H is uniquely determined by β:

(23) $\qquad H(t) = \dfrac{(\tan \beta(t))'}{\tan \beta(t)}$.

If β is a solution of (20) which satisfies (21), then the proof of X (1.23) shows that

(24) $\qquad \lim_{t \to +\infty} H(t) = k_v \quad \text{and} \quad \lim_{t \to -\infty} H(t) = k_u$

where k_u, k_v are the respective polynomial degrees of the eigenmaps u, v.

Direct substitution of (22) into (20) shows that

(25) $\qquad H'(t) + H^2(t) + D(t)H(t) + G(t) > 0 \quad$ for all $\quad t \in \mathbf{R}$.

As in X (1.28–30), the inequality (25) determines an area HND in which the function H does not decrease. Now suppose that $(DC)_1$ does not hold. Straightforward computations lead to the qualitative description of HND in Figure 15:

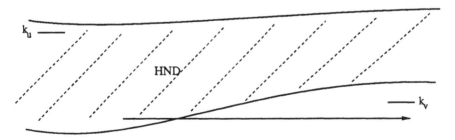

Figure 15

Thus it is clear from (24) that there is no solution. Conversely, if $(DC)_1$ holds then HND behaves qualitatively as in Figure 16.

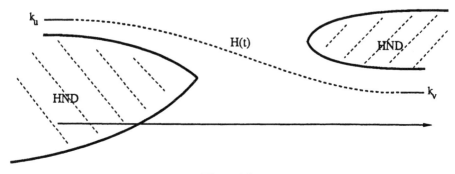

Figure 16

In this case we can construct (see [PR]) explicit subsolutions and supersolutions of (20), using functions $H(t)$ behaving qualitatively as in Figure 16.

The existence of a harmonic join follows from a comparison argument of the sort illustrated in the first part of this Appendix.

The proof of X (1.15) occurs in [R4], where the idea of using the area HND was first introduced; see also [PR].

REFERENCES

[Ab1] U. Abresch, *Isoparametric hypersurfaces with four or six principal curvatures*, Math. Ann. **264** (1983) 283–302.

[Ab2] U. Abresch, *Constant mean curvature tori in terms of elliptic functions*, J. Reine Angew. Math. **374** (1987) 169–192.

[A1] J.F. Adams, *On the nonexistence of elements of Hopf invariant one*, Ann. Math. **72** (1960) 20–104.

[A2] J.F. Adams, *On the groups $J(X)$*, Topology **3** (1965) 137–171; **5** (1966) 21–71.

[Ad] J. Adem, *Construction of some normed maps*, Bol. Soc. Mat. Mex **20** (1975) 59–75.

[Al] A.D. Alexandrov, *Uniqueness theorems for surfaces in the large*, V. Amer. Math. Soc. Transl. **21** (1962) 412–416.

[Alm] F.J. Almgren, *Some interior regularity theorems for minimal surfaces and an extension of Bernstein's theorem*, Ann. Math. **84** (1966) 277–292.

[ABS] M.F. Atiyah, R. Bott and A. Shapiro, *Clifford modules*, Topology 3 Suppl. **1** (1964) 3–38.

[Ba] P. Baird, *Harmonic maps with symmetry, harmonic morphisms and deformations of metrics*, Research Notes in Math. 87, Pitman (1983).

[BE] P. Baird and J. Eells, *A conservation law for harmonic maps*, Springer Lecture Notes 894 (1981) 1–25.

[BW] P. Baird and J.C. Wood, *Harmonic morphisms, Seifert fibre spaces and conformal foliations*, Proc. London Math. Soc. **64** (1992) 170–196.

[BdC] J.L. Barbosa and M. do Carmo, *Stability of hypersurfaces with constant mean curvature*, Math. Z. **185** (1984) 339–353.

[BdCE] J.L. Barbosa, M. do Carmo and J. Eschenburg, *Stability of hypersurfaces of constant mean curvature in Riemannian manifolds*, Math. Z. **197** (1988) 123–138.

[B] P. Baum, *Quadratic maps and stable homotopy groups of spheres*, Ill. J. Math. **11** (1967) 586–595.

[BB] L. Bérard Bergery and J-P. Bourguignon, *Laplacians and Riemannian submersions with totally geodesic fibres*, Ill. J. Math. **26** (1982) 181–200.

[BGM] M. Berger, P. Gauduchon and E. Mazet, *Le spectre d'une variété riemannienne*, Springer Notes **194** (1971).

[BV] M-F. Bidaut-Veron, *Global existence and uniqueness results for singular solutions of the capillarity equation*, Pac. J. Math. **125** (1986) 317-333.

[Bo] A.I. Bobenko, *All constant mean curvature tori in R^3, S^3, H^3 in terms of theta-functions*, Math. Ann. **290** (1991) 209–245.

[BK] J. Bochnak and W. Kucharz, *Realization of homotopy classes by algebraic mappings*, J. Reine Angew. Math. **377** (1987) 159–169.

[BDG] E. Bombieri, E. De Giorgi and E. Giusti, *Minimal cones and the Bernstein problem*, Inv. Math. **7** (1969) 243–268.

[Ca] E. Calabi, *Minimal immersions of surfaces in Euclidean spheres*, J. Diff. Geo. **1** (1967) 111–125.

[dCD] M. do Carmo and M. Dajczer, *Rotation hypersurfaces in spaces of constant curvature*, T.A.M.S. **227** (1983) 685–707.

[dCL] M. do Carmo and H.B. Lawson, *On Alexandrov–Bernstein theorems in hyperbolic space*, Duke Math. J. **50** (1983) 995–1003.

[dCW] M. do Carmo and N. Wallach, *Minimal immersions of spheres into spheres*, Ann. Math. **93** (1971) 43–62.

[C1] E. Cartan, *Familles de surfaces isoparamétriques dans les espaces à courbure constante*, Ann. Mat. Pura Appl. **17** (1938) 177–191.

[C2] E. Cartan, *Sur des familles remarquables d'hypersurfaces isoparamétriques dans les espaces sphériques*, Math. Z. **45** (1939) 335–367.

[C3] E. Cartan, *Sur des familles remarquables d'hypersurfaces isoparamétriques des espaces sphériques à 5 et 9 dimensions*, Revista Univ. Tucuman. Sér. A. **1** (1940) 5–22.

[CW] S. Carter and A. West, *Isoparametric systems and transnormality*, Proc. London Math. Soc (3) **51** (1985) 520–542.

[C] J.W.S. Cassels, *On the representation of rational functions as sums of squares*, Acta Arith. **9** (1964) 79–82.

[Ch]   K–C. Chang, *Infinite dimensional Morse theory and its applications*, Sém. Math. Sup. Montréal **97** (1985).

[CE]   K–C. Chang and J. Eells, *Unstable minimal surface coboundaries*, Acta Math. Sinica **2** (1986) 233–247.

[Cha] B. Charlet, *Le problème de Bernstein sphérique*, Exp XI, Théorie des variétés minimales et applications, Astérisque **154-5** (1987) 221–243.

[XPC] X–P. Chen, *Harmonic mapping and Gauss mapping*, Proc. 1981 Shanghai-Hefei Symp. Diff. Geo. Diff. Eq., Sci. Press, Beijing (1984) 51–53.

[Che] S–S. Chern, *Minimal surfaces in an euclidean space of N dimensions*, Symp. M. Morse, Princeton (1965) 187–198.

[CCK] S-S. Chern, M. do Carmo and S. Kobayashi, *Minimal submanifolds of a sphere with second fundamental form of constant length*, Fun. Analysis and Related Fields. Springer (1970) 59–75.

[CL]   E. Coddington and N. Levinson, *Theory of ordinary differential equations*, McGraw–Hill (1955).

[CF1] P. Concus and R. Finn, *On capillary free surfaces in the absence of gravity*, Acta Math. **132** (1974) 177-198. *On capillary free surfaces in a gravitational field*, Acta Math. **132** (1974) 207–223.

[CF2] P. Concus and R. Finn, *A singular solution of the capillary equation I,II*, Inv. Math. **29** (1975) 143-148 and 149-159.

[De]   C. Delaunay, *Sur la surface de révolution dont la courbure moyenne est constante*, J. Math. Pures et Appl. **1** (1841) 309–320.

[Di]   W–Y. Ding, *Symmetric harmonic maps between spheres*, Comm. Math. Phys. **118** (1988) 641–649.

[E]   B. Eckmann, *Gruppentheoretischer Beweis des Satzes von Hurwitz-Radon über die Komposition quadratischer Formen*, Comm. Math. Helv. **15** (1943) 358-366.

[Ee]   J. Eells, *On the surfaces of Delaunay and their Gauss maps*, Proc. IV Inter. Colloq. Diff. Geom. Santiago de Compostela (1978) 97-116. Reprinted in Math. Intell. **9** (1987) 53-57.

[EL1] J. Eells and L. Lemaire, *A report on harmonic maps*, Bull. London Math. Soc. **10** (1978) 1–68.

[EL2] J. Eells and L. Lemaire, *Another report on harmonic maps*, Bull. London Math. Soc. **20** (1988) 385–524.

[EL3] J. Eells and L. Lemaire, *Selected topics in harmonic maps*, CBMS. Regional Conf. Series 50, A.M.S. Providence (1983).

[EL4] J. Eells and L. Lemaire, *Examples of harmonic maps from disks to hemispheres*, Math. Z. **185** (1984) 517-519.

[ER] J. Eells and A. Ratto, *Harmonic maps between spheres and ellipsoids*, Inter. J. Math. **1** (1990) 1–27.

[ES] J. Eells and J.H. Sampson, *Harmonic mappings of Riemannian manifolds*, Amer. J. Math. **86** (1964) 109–160.

[ElS] A. El Soufi, *Immersions minimales et applications harmoniques d'une variété compacte dans les sphères*, Preprint Univ. Tours (1990).

[EKT] N. Ercolani, H. Knörrer, and E. Trubowitz, *Hyperelliptic curves that generate constant mean curvature tori in $\mathbb{R}^3$*, Preprint (1991).

[F] D. Ferus, *Isometric immersions of constant curvature manifolds*, Math. Ann. **217** (1975) 155–156.

[FK] D. Ferus and H. Karcher, *Non-rotational minimal spheres and minimizing cones*, Comm. Math. Helv. **60** (1985) 247–269.

[FP] D. Ferus and U. Pinkall, *Constant curvature 2-spheres in the 4–sphere*, Math. Z. **200** (1989) 265–271.

[Fu] B. Fuglede, *Harmonic morphisms between Riemannian manifolds*, Ann. Inst. Fourier **28** (1978) 107–144.

[Gi] G. Gigante, *A note on harmonic morphisms*, Preprint Univ. Camerino (1983).

[Gc] E. Giusti, *Minimal surfaces and functions of bounded variation*, Birkhauser (1984).

[GWY] H. Gluck, F. Warner and C-T. Yang, *Division algebras, fibrations of spheres by great spheres and the topological determination of space by the gross behaviour of its geodesics*, Duke Math. J. **50** (1983) 1041–1076.

[Gu] M.A. Guest, *The energy function and homogeneous harmonic maps*, Proc. London Math. Soc. **62** (1991) 77-98.

- [G] R. Gulliver, *Necessary conditions for submanifolds and currents with prescribed mean curvature vector*, Ann. Math. Studies **103** (1983) 225–242.

- [GOR] R. Gulliver, R. Osserman and H. Royden, *A theory of branched immersions of surfaces*, Amer. J. Math. **95** (1973) 750–812.

- [HN] J. Hano and K. Nomizu, *Surfaces of revolution with constant mean curvature in Lorentz–Minkowski space*, Tôhoku Math. J. **36** (1984) 427-437.

- [Ha] P. Hartman, *Ordinary differential equations*, Wiley (1964).

- [He] H. Hefter, *Dehnungsuntersuchungen an Sphärenabbildungen*, Inv. Math. **66** (1982) 1–10.

- [Her] R. Hermann, *On geodesics that are also orbits*, Bull. Amer. Math. Soc. **66** (1960) 91-93.

- [Hi] N.J. Hitchin, *Harmonic maps from a 2-torus to the 3-sphere*, J. Diff. Geo. **31** (1990) 627-710.

- [Hof] D. Hoffman, *Surfaces of constant mean curvature in manifolds of constant curvature*, J. Diff. Geo. **8** (1973) 161–176.

- [HoO] D. Hoffman and R. Osserman, *The geometry of the generalized Gauss map*, Mem. Amer. Math. Soc. **236** (1980).

- [Ho1] H. Hopf, *Uber Flächen mit einer Relation zwischen den Hauptkrümmungen*, Math. Nach. **4** (1950/1) 232–249.

- [Ho2] E. Hopf, *Elementare Bemerkungen über die Lösungen partieller Differentialgleichungen zweiter Ordnung von elliptishen Typus*, Sitz. Ber. Preuss Akad. **19** (1927) 147–152.

- [HW] R. Howard and S.W. Wei, *Nonexistence of stable harmonic maps to and from certain homogeneous spaces and submanifolds of Euclidean space*, Trans. Amer. Math. Soc. **294** (1986) 319-331.

- [HH] W-T. Hsiang and W-Y. Hsiang, *On the existence of codimension–one minimal spheres in compact symmetric spaces of rank 2. II*, J. Diff. Geo. **17** (1982) 583–594.

- [H1] W-Y. Hsiang, *On the compact homogeneous minimal submanifolds*, Proc. Nat. Acad. Sci. **56** (1966) 5–6.

- [H2] W-Y. Hsiang, *Minimal cones and the spherical Bernstein problem, I*, Ann. Math. **118** (1983) 61–73; and *II*, Inv. Math. **74** (1983) 351-369.

[H3] W-Y. Hsiang, *On generalization of theorems of A.D. Alexandrov and C. Delaunay on hypersurfaces of constant mean curvature*, Duke Math. J. **49** (1982) 485–496.

[H4] W-Y. Hsiang, *Generalized rotational hypersurfaces of constant mean curvature in the Euclidean spaces. I*, J. Diff. Geo. **17** (1982) 337-356.

[HL] W-Y. Hsiang and H.B. Lawson, *Minimal submanifolds of low cohomogeneity*, J. Diff. Geo. **5** (1971) 1–38. Corrections in F. Uchida, *An orthogonal transformation group of* $(8k - 1)$-*sphere*, J. Diff. Geo. **15** (1980) 569–574.

[HS] W-Y. Hsiang and I. Sterling, *On the construction of non–equatorial hypersurfaces in $S^n(1)$ with stable cones in $\mathbb{R}^{n+1}$*, Proc. Nat. Acad. Sci. **81** (1984) 8035–8036.

[HT] W-Y. Hsiang and P. Tomter, *On minimal immersions of S^{n-1} into $S^n(1)$*, $n \geq 4$, Ann. Sci. ENS **20** (1987) 201–214.

[HTY] W-Y. Hsiang, Z-H. Teng and W-C. Yu, *New examples of constant mean curvature immersions of $(2k - 1)$-spheres into Euclidean $2k$-space*, Ann. Math. **117** (1983) 609–625.

[HY] W-Y. Hsiang and W-C. Yu, *A generalization of a theorem of Delaunay*, J. Diff. Geo. **16** (1981) 161–177.

[I1] T. Ishihara, *A mapping of Riemannian manifolds which preserves harmonic functions*, J. Math. Kyoto Univ. **19** (1979) 215–229.

[I2] T. Ishihara, *The harmonic Gauss maps in a generalized sense*, J. London Math. Soc. **26** (1982) 104-112.

[Ka] N. Kapouleas, *Compact constant mean curvature surfaces in Euclidean three-space*, J. Diff. Geo. **33** (1991) 683–715.

[KPS] H. Karcher, U. Pinkall and I. Sterling, *New minimal surfaces in S^3*, J. Diff. Geo. **28** (1988) 169–185.

[KW] H. Karcher and J.C. Wood, *Non-existence results and growth properties for harmonic maps and forms*, J. Reine Angew. Math. **353** (1984) 165-180.

[KN] S. Kobayashi and K. Nomizu, *Foundations of differential geometry*, I, II, Interscience (1963) and (1969).

[LU] O. Ladyzenskaya and N. Ural'ceva, *Linear and quasilinear elliptic equations*, Academic Press (1968).

[La1] K-Y. Lam, *Construction of nonsingular bilinear maps*, Topology **6** (1967) 423-426.

[La2] K-Y. Lam, *Some interesting examples of nonsingular bilinear maps*, Topology **16** (1977) 185–188.

[La3] K-Y. Lam, *Some new results in compositions of quadratic forms*, Inv. Math. **79** (1985) 467–474.

[L1] H.B. Lawson, *Local rigidity theorems for minimal hypersurfaces*, Ann. Math. **89** (1969) 187-197.

[L2] H.B. Lawson, *Complete minimal surfaces in S^3*, Ann. Math. **92** (1970) 335–374.

[L3] H.B. Lawson, *Lectures on minimal submanifolds*, Publish or Perish (1980).

[Le] L. Lemaire, *Applications harmoniques de surfaces riemanniennes*, J. Diff. Geo. **13** (1978) 51–78.

[Leu] P.F. Leung, *A note on stable harmonic maps*, J. London Math. Soc. **29** (1984) 380-384.

[Lo] J.L. Loday, *Applications algébriques du tore dans la sphère et de $S^p \times S^q$ dans S^{p+q}*, Springer Notes No. 342 (1973) 79-91.

[M] Y. Matsushima, *Vector bundle valued harmonic forms and immersions of Riemannian manifolds*, Osaka J. Math. **8** (1971) 1–13.

[Ma] E. Mazet, *La formule de la variation seconde de l'énergie au voisinage d'une application harmonique*, J. Diff. Geo. **8** (1973) 279-296.

[Mi] J. Milnor, *Curvature of left invariant metrics on Lie groups*, Adv. Math. **21** (1976) 293–329.

[MSY] D. Montgomery, H. Samelson and C-T. Yang, *Exceptional orbits of highest dimension*, Ann. Math. **64** (1956) 131–141.

[Moo] J.D. Moore, *Isometric immersions of space forms in space forms*, Pac. J. Math. **40** (1972) 157-166.

[Mo] C.B. Morrey, *Multiple integrals in the calculus of variations*, Grundlehren Band **130** Springer (1966).

[Mü] H.F. Münzner, *Isoparametrische Hyperflächen in Sphären*, Math. Ann. **251** (1980) 57-71; **256** (1981) 215-232.

[N] K. Nomizu, *Elie Cartan's work on isoparametric families of hypersurfaces*, Proc. Symp. Pure Math. A.M.S. 27 (1975) 191–200.

[O] Y. Ohnita, *Stability of harmonic maps and standard minimal immersions*, Tôhoku Math. J. 38 (1986) 259-267.

[ON] B. O'Neill, *The fundamental equations of a submersion*, Mich. Math. J. 13 (1966) 459–469.

[Ot] T. Otsuki, *Minimal hypersurfaces in a Riemannian manifold of constant curvature*, Amer. J. Math. 92 (1970) 145–173.

[P] R. Palais, *The principle of symmetric criticality*, Comm. Math. Phys. 69 (1979) 19–30.

[PT] R. Palais and C-L. Terng, *Reduction of variables for minimal submanifolds*, Proc. A.M.S. 98 (1986) 480–484.

[Pa] M. Parker, *Orthognal multiplications in small dimensions*, Bull. L.M.S. 15 (1983) 368–372.

[PR] V. Pettinati and A. Ratto, *Existence and non–existence results for harmonic maps between spheres*, Ann. SNS Pisa, Ser.IV, 17 (1990) 273–282.

[PS] U. Pinkall and I. Sterling, *On the classification of constant mean curvature tori*, Ann. Math. 130 (1989) 407-451.

[Pf] A. Pfister, *Multiplikative quadratische Formen*, Arch. Math. 16 (1965) 363–370.

[Pl1] A.I. Pluzhnikov, *Harmonic mappings of Riemann surfaces and foliated manifolds*, Mat. Sb. (N.S.) 113 (1980) No.7(90) 339–347, 352. English translation Math. USSR Sb. 41 (1982) 281–287.

[Pl2] A.I. Pluzhnikov, *A topological criterion for the attainability of global minima of an energy function*, Nov. Glob. Anal. Voronizh. Gos. Univ. 177 (1986) 149-155.

[Q] D. Quillen, *The Adams conjecture*, Topology 10 (1970) 67–80.

[R1] A. Ratto, *Harmonic maps of spheres and the Hopf construction*, Topology 28 (1989) 379–388.

[R2] A. Ratto, *Harmonic maps from deformed spheres to spheres*, Amer. J. Math. 111 (1989) 225–238.

[R3] A. Ratto, *On harmonic maps between S^3 and S^2 of prescribed Hopf invariant*, Math. Proc. Camb. Phil. Soc. **104** (1988) 273–276.

[R4] A. Ratto, *Harmonic maps of spheres and equivariant theory*, Warwick Thesis (1987).

[Ra] J. Rawnsley, *Equivariant harmonic maps*, Preprint Univ. Warwick (1984).

[Re] R. Reilly, *Applications of the Hessian operator in a Riemannian manifold*, Ind. Math. J. **26** (1977) 459–472.

[R] E.A. Ruh, *Minimal immersions of 2-spheres in S^4*, Proc. Am. Math. Soc. **28** (1971) 219-222.

[RV] E. Ruh and J. Vilms, *The tension field of the Gauss map*, Trans. A.M.S. **149** (1970) 569–573.

[SU1] R. Schoen and K. Uhlenbeck, *A regularity theory for harmonic maps*, J. Diff. Geo. **17** (1982) 307–335 and **18** (1983) 329.

[SU2] R. Schoen and K. Uhlenbeck, *Regularity of minimizing harmonic maps into the sphere*, Inv. Math. **78** (1984) 89–100.

[Sc] G.W. Schwarz, *Smooth functions invariant under the action of a compact Lie group*, Topology **14** (1975) 63–68.

[Se] H. Sealey, *Some properties of harmonic mappings*, Warwick Thesis (1980).

[Si] J. Simons, *Minimal varieties in Riemannian manifolds*, Ann. Math. **88** (1968) 62-105.

[Sm1] R.T. Smith, *Harmonic mappings of spheres*, Warwick Thesis (1972).

[Sm2] R.T. Smith, *Harmonic mappings of spheres*, Amer. J. Math. **97** (1975) 364–385.

[Sm3] R.T. Smith, *The second variation formula for harmonic mappings*, Proc. Amer. Math. Soc. **47** (1975) 229-236.

[Smy1] B. Smyth, *Submanifolds of constant mean curvature*, Math. Ann. **205** (1973) 265–280.

[Smy2] B. Smyth, *The geometry of bounded solutions of the sinh–Gordon equation*, (Unpublished).

[Sp] M. Spivak, *A comprehensive introduction to differential geometry*, Vol.4, Addendum 3, Publish or Perish (1970).

[St] M. Struwe, *On a critical point theory for minimal surfaces spanning a wire in R^n*, J. Reine Angew. Math. **349** (1984) 1–23.

[Ta] T. Takahashi, *Minimal immersions of Riemannian manifolds*, J. Math. Soc. Japan **18** (1966) 380–385.

[Te] C–L. Terng, *Isoparametric submanifolds and their Coxeter groups*, J. Diff. Geo. **21** (1985) 79–107.

[To] P. Tomter, *The spherical Bernstein problem in even dimensions and related problems*, Acta Math. **158** (1987) 189–212.

[Tot1] G. Toth, *Harmonic and minimal maps with applications in geometry and physics*, E. Horwood (1984).

[Tot2] G. Toth, *Harmonic maps and minimal immersions through representation theory*, Persp. in Math. Academic Press (1990).

[TotA] G. Toth and G. D'Ambra, *Parameter space for harmonic maps of constant energy density into spheres*, Geo. Ded. **17** (1984) 61-67.

[Tr] A. Treibergs, *Entire spacelike hypersurfaces of constant mean curvature in Minkowski space*, Inv. Math. **66** (1982) 39–56.

[Ty] A.V. Tyrin, *Critical points of the multidimensional Dirichlet functional*, Mat. Sb. **124** (166) (1984) 146-158. English translation: Math. USSR Sb. **52** (1985) 141-153.

[U1] H. Urakawa, *The first eigenvalue of the Laplacian for a positively curved homogeneous Riemannian manifold*, Comp. Math. **59** (1986) 57-71.

[U2] H. Urakawa, *Stability of harmonic maps and eigenvalues of the Laplacian*, Trans. Amer. Math. Soc. **301** (1987) 557-589.

[Va] A.J. Vanderwinden, *Examples of harmonic mappings obtained by reduction*, Preprint Univ. Bruxelles (1991).

[V] J. Vilms, *Totally geodesic maps*, J. Diff. Geo. **4** (1970) 73–79.

[W] N. Wallach, *Minimal immersions of symmetric spaces into spheres*, Dekker (1972) 1–40.

[Wa] Q–M. Wang, *Isoparametric maps of Riemannian manifolds and their applications*, Adv. in Sci. of China Math. **2** (1987) 79-103.

[Wt] B. Watson, *The first Betti numbers of certain locally trivial fibre spaces*, Bull. A.M.S. **78** (1972) 392–393.

[We1] H. Wente, *Counterexample to a conjecture of H. Hopf*, Pac. J. Math. **121** (1986) 193–243.

[We2] H. Wente, *Twisted tori of constant mean curvature in* $\mathbf{R}^3$, *Non-linear P.D.E.*, Sem. on new results, Aspects of Math. **E10**, Bonn (1986).

[We] A. West, *Isoparametric systems*, Geo. and Top. of Subman. Luminy (1987), World Scientific Press.

[Wo1] R. Wood, *A note on harmonic polynomial maps*, (Unpublished).

[Wo2] R. Wood *Polynomial maps from spheres to spheres*, Inv. Math. **5** (1968) 163–168.

[X1] Y-L. Xin, *Some results on stable harmonic maps*, Duke Math. J. **47** (1980) 609-613.

[X2] Y-L. Xin, *On harmonic representative of* $\pi_{2m+1}(S^{2m+1})$, Preprint Fudan Univ. (1991).

[Y] P. Yiu, *Quadratic forms between spheres and the non-existence of sums of squares formulae*, Math. Proc. Comb. Phil. Soc. **100** (1986) 493–504.

INDEX

Adapted frame IV (1.11)

Basic tension field IV (3.8)

Basic vector field IV (1.3)

Berger metric App. 1 (38)

Bi-eigenmap VIII (2.11)

Capillarity equation App. 3(5)

Clifford torus V (2.8)

Codifferential IV (2.5)

Cohomogenity IV (3.18)

Conformal (weakly conformal) III (1.5)

Constant mean curvature II (1.2)

Critical point of E I (1.8)

Damping conditions IX (4.1), X (1.6)

Delaunay surface V (3.14)

Dilatation (ellipsoidal IX (1.1)

Eiconal VIII (1.19)

Eigenmap VIII (1.1), (1.11), (1.30)

Einstein manifold App. 1 (26)

Ellipsoidal α-join IX (1.5)

Energy density of φ I (1.3)

Energy of φ I (1.4)

Equivariant map V (1.2)

Exceptional orbit IV (3.15)

First fundamental form of φ I (1.2)

Gauss map II (1.6)

Graph App. 3 (1)

Grassmann manifold II (1.5)

Harmonic map I (1.7)

Harmonic morphism IV (2.1)

E-Hessian App. 1 (1), V-Hessian App. 1 (40)

Hopf construction VIII (2.3), (2.16), α-Hopf construction X (1.1)

Hopf fibration VIII (2.8)

Hopf form VIII (3.18)

Horizontally conformal IV (2.1)

Horizontal map IV (4.1)

E-index App. 1 (6), V-index App. 1 (44).

Isoparametric function VIII (1.19)

Isoparametric map IV (3.4)

Jacobi field App. 1 (6)

Jacobi operator App. 1 (6)

J-homomorphism X (2.3)

Join V (2.12), IX (1.1)

α-Join IX (1.6)

Laplacian I (1.7)

Lines of curvature III (4.1)

Mean curvature II (1.1)

Minimal cone VIII (1.14)

Minimal graph App. 3 (1)

Minimal immersion I (2.3)

Mountain pass lemma IX (2.11)

Nodoid V (3.15)

c-nullity App. 2 (1)

Nullity App. 1 (6)

Obata's Gauss map II (1.11)

Orbit type IV (3.15)

Orthogonal multiplication VIII (2.1)

Palais-Smale conditions IX (2.7)

Parallel mean curvature II (1.2)

Pauli spin matrices V (4.3)

Pendulum type equation IX (1.11), X (1.5), App. 4

Polynomial map VIII (3.1)

Prime integral V (3.7)

Principal orbit type IV (3.15)

Quadratic map of spheres VIII (3.17), (3.19), X (2.12)

Reduced energy functional IX (1.13)

Regular cone VIII (1.14)

Regular fibre IV (3.2)

Relative nullity App. 2(1)

Riemannian immersion I (1.15)

Riemannian submersion IV (1.2)

Scalar curvature App. 1 (26)

Second fundamental form of φ I (1.5).

E-Stable App. 1 (6)

Sinh-Gordon equation III (3.2)

Spectrum of Δ^{S^m} VIII (1.9)

Stable J-homomorphism V (2.6)

Standard minimal immersion VIII (1.17), VIII (4.2)

Stiefel manifolds VIII (1.25)

Tension field I (1.7)

Third fundamental form App. 2 (5)

Totally geodesic map I (1.6)

Totally umbilic III (1.16)

Transnormal map IV (3.9)

Umbilic points III (1.16)

Unduloid V (3.15)

Variations of φ I (1.1)

Vector field along φ I (1.7)

Veronese map VIII (1.17)

Veronese surface I (3.6)

Winding number VII (1.11)